Beyond Human Limits: The Productivity Singularity

Harnessing AI to Work Smarter, Live Better, and Grow Your Wealth - Without Losing What Makes You Human

Anthony I. Case

Beyond Human Limits: The Productivity Singularity

Harnessing AI to Work Smarter, Live Better, and Grow Your Wealth - Without Losing What Makes You Human

For permission requests, please contact the publisher at:

anthonycasebooks@gmail.com

ISBN: 979-8-234-17396-6

Cover design by Anthony Case

Printed in the United States of America

“For my family- my greatest inspiration and the reason I believe the future is always worth building.”

Contents

Introduction: Crossing the Threshold 1

Part I — The New Frontier 9

1. The Age of the Productivity Singularity 10
2. Breaking Past Human Limits 21
3. Without Losing Your Humanity 32

Part II — Productivity Reimagined 42

4. Automating the Mundane 43
5. Decision-Making in the AI Era 50
6. Reprogramming the Mindset for Exponential Growth 60

Part III — Health and Energy as the New Wealth 69

7. AI for Health Habits and Recovery 70
8. Protecting Attention and Cognitive Energy 80
9. Emotional Intelligence in a Digital Age 89

Part IV — AI in Business 99

10. The AI Advantage in Business and Career 100

11. Wealth Creation in the AI Economy 112
12. The Democratization of Intelligence 126
Part V — AI in Everyday Life 137
13. AI for Creativity 138
14. AI in the Home 149
15. AI for Lifelong Learning and Hobbies 160
Part VI — Living Beyond Human Limits 171
16. Designing Your AI-Integrated Life 172
17. The Ethics of AI and Human Choice 184
18. What's Next in AI 197
Conclusion: Beyond Human Limits—Your Next Advantage 211
AI Toolkit: Practical Tools for Every Area of Life 220
Recommended AI Starter Stack 227
About the Author 230

Introduction: Crossing the Threshold

The world no longer runs on effort alone. It runs on **amplification**.

You can feel that truth before you know how to explain it. The first ping of the day lands like a starter pistol. Before your feet hit the floor, the race has already begun. Messages wait. Calendars tighten. Small fires flicker at the edges of your attention. By night, the house is quiet, but your mind isn't. It's still sorting threads, replaying conversations, building tomorrow before today has fully ended.

And somewhere in the middle of all that noise, a thought begins to form. The hardest workers aren't always the ones getting the most done.

You know the people I mean. The ones who seem steady instead of frantic. The ones whose projects ship on time without a trail of drama. The ones who don't look rushed, yet somehow move faster than everyone else. They aren't better than you. They aren't superhuman. They aren't secretly working twice as many hours. More and more, **they're simply using tools that multiply what they do best.**

That shift changes everything.

Your phone is no longer just a phone. It's a studio, a research assistant, a note-taker that never forgets, a coach that doesn't get tired, a partner in the first draft, a system for catching what your overloaded brain would otherwise drop. Intelligence is no longer locked away in distant systems, accessible only to specialists or giant companies. It's moved into your pocket. It's here, now, waiting for direction.

This book is about learning how to direct it.

Not in some abstract, futuristic sense. Not as a science-fiction fantasy. Not as a vague promise that one day technology will change your life. It already has. The question is whether you'll use it deliberately, or whether it'll simply shape your days without your consent.

Because this isn't really a book about tools. It's a book about leverage. It's about what happens when your intent meets on-demand capability. It's about the moment when one hour begins to work like several, when one clear decision opens space in an entire week, when one person begins to operate at something closer to team scale without becoming colder, flatter, or less human in the process.

That's the shift I call the Productivity Singularity.

Not a robot uprising. Not some distant economic theory. A present-tense change in the way work, learning, creating, leading, and living now happen. A bend in the curve of results. A new relationship between what you want to do and what you're actually able to get done.

And the best part is this: it doesn't require you to become someone else.

A Better Way to Carry the Week

Picture a kitchen table on a Tuesday night. Two kids are trading homework questions across the table. A dog thumps its tail beneath a chair. The day has already taken more out of you than you meant to give, and now you open your calendar and feel that familiar knot in your stomach. There are three obligations and one usable block of time. Dinner is still unresolved. Pickup is hanging over the evening. The week feels like it's beginning to tilt.

The old pattern is familiar. You juggle. You apologize. You move things badly instead of thoughtfully. You absorb the friction into your own body and call it responsibility. You get through the week, but not cleanly. And when it's over, you don't feel proud. You feel spent.

Now imagine something different.

Instead of bracing, you pause. You turn to your AI copilot and ask a calm, specific question:

"Help me reshape this week to protect two deep-work blocks, handle the kid pickup, and plan dinner without chaos."

Ten minutes later, the week looks different. A call has been moved. A focus block is protected. A shared grocery list has been created. Three easy meals are in rotation. What felt jammed now has shape. The knot loosens.

Nothing magical happened. No genie appeared. No one handed you a second life or a longer day.

You simply aimed leverage.

The same thing happens in smaller, quieter moments that matter just as much. Late one afternoon, a manager sits staring at an empty email window, trying to find the right tone for a delicate update. The message has to be clear without being cold, honest without being clumsy, kind without drifting into vagueness. The right words are there somewhere, but they're just out of reach. So instead of spending fifty drained minutes circling the same three sentences, the manager types a prompt:

"Draft a clear, kind message with two options and a next step, in my voice."

A draft appears. Not perfect, but close. A sentence is trimmed. A personal line is added. The judgment remains human. The decision remains human. But the path is shorter now. Five minutes, not fifty.

And then there's the kind of moment that changes more than a day.

A weekend finally opens up enough to let an old idea breathe. For months, maybe longer, a simple product has been sitting in the back of your mind—a template, a guide, something small that could genuinely help people. In another season of life, you'd have overthought it into oblivion. You'd have waited for the perfect plan, the perfect branding, the perfect certainty that never actually comes. But now you move. You sketch an outline with an AI assistant. You polish the layout. You make a lightweight cover. You publish the page.

By Monday morning, a message arrives from someone you've never met: This saved me an hour. Thank you.

It's a small thing. **But small things compound**. A recovered hour becomes a repeated hour. A simple product becomes proof. A proof point becomes confidence. Confidence becomes motion. And motion, over time, becomes a very different life.

This is the promise of the age you're living in. Not ease without effort, but effort that finally scales. Not hustle for its own sake, but output with direction. Not doing more for applause, but making room for what actually matters.

The Rhythm of a New Kind of Productivity

There are a few things that won't change inside these pages.

You stay in charge. Tools propose; you decide. People come before throughput. Health remains non-negotiable, because energy is the engine of output. And ethics stays in the loop, because the goal is to augment judgment, not outsource it.

That's why this book isn't built on techno-mysticism or breathless hype. You won't find twenty-step morning rituals designed for people with no children, no meetings, and unlimited control over their days. You won't find buzzword soup. What you'll find is a practical way to work with the world as it is now: plain language, repeatable rhythms, and useful experiments you can run between meetings, after bedtime, before school pickup, or in the margins of an already full life.

At the heart of the book is a simple operating rhythm.

First, you learn to spot the leaks. You begin to notice where your time and energy quietly drain away—notifications that peck at your focus, meetings that don't need your mind, tasks that repeat so often they've become invisible. The moment you can name those leaks, they stop feeling like fog. They become fixtures. And fixtures can be moved.

Next, you create a baseline. You teach the tools to take the first pass: the formatting, the checklists, the rough drafts, the handoffs, the summaries, the recurring setup work that wears you down before the meaningful work even begins. A baseline doesn't replace you. It meets you halfway. It gives you something to react to instead of a blank page, and that alone can change the shape of a day.

Then you reserve the human work on purpose. Judgment. Story. Relationships. Strategy. Discernment. Taste. Those don't get pushed to the margins. They become the center. You guard them like prime real estate because they are. You spend your best energy there.

Finally, you run short reviews. You keep what helps. You delete what doesn't. You level up weekly instead of yearly. The rhythm is light, but it's relentless: What got easier? What still drags? What will I try next?

This rhythm works in more places than you might expect. It shows up at work, obviously, but also in your health, your finances, your home, your learning, your creativity, your leadership, and your relationships. It works because leverage isn't only a professional concept. It's a life concept.

But leverage without boundaries is just another form of noise.

That's why this book also asks you to decide where not to optimize. You'll choose your AI-free zones. Family dinners. Quiet mornings. Bedtime rituals. Long walks. Deep conversations. The moments where speed isn't the goal, and presence is. **Power without restraint only creates more static. Power with values creates breathing room.**

And that breathing room matters more than most people realize. Because what you're really after isn't just efficiency. It isn't just speed. It isn't even just success. It's a life where your week finally fits your values. Where your work lands heavier because your mind feels lighter. Where your wealth can grow without your health quietly shrinking. Where your family gets more of the version of you that's actually there. Where you can build, lead, serve, and create without disappearing inside the process.

That's what this book is for.

It's a primer for people with constraints and ambition. For builders. For parents. For leaders. For learners. For people who aren't waiting around for perfect conditions but are no longer willing to grind themselves into the floor just to prove they care.

You don't need more hours.

You need better leverage. And you need a way to point it.

So here's the invitation.

You can keep grinding through noise, or you can design a week that protects what only you can do. You can let screens siphon your attention, or you can direct them to return it. You can wait for certainty, or you can run small experiments that compound into confidence.

Turn the page.

Let's make your days feel lighter and your work land heavier. Let's grow your wealth without shrinking your health. Let's stay unmistakably human while learning to operate beyond the limits that used to hold us back.

The threshold is right here.

Step through

Part I – The New Frontier

"New discoveries in science and their flow of new inventions will continue to create a thousand new frontiers for those who still would adventure."

– Herbert Hoover

1

The Age of the Productivity Singularity

The text on Alex's screen blurred into a gray river—thirty-two emails unread since noon, a calendar grid stacked like Tetris, and a Slack ping that said, "Quick question?" which never meant quick. From the kitchen came the sound of a drawer closing and a voice—"Hey, who's doing the Thursday pickup?"—and Alex felt that familiar tug-of-war: be a reliable teammate at work, be a reliable human at home, and try not to disappoint either.

Alex dragged the cursor toward the calendar to start moving blocks around, then paused. A week ago, a friend had said, just try using AI once when you feel stuck. Ask the machine to do the first pass.

So Alex opened Copilot and pasted the mess: meetings, deadlines, the soccer schedule, the dentist appointment that kept migrating week to week.

> ***"Protect two 90-minute deep-work blocks, keep the pick-ups, and draft the three reschedule messages I'll need to send. Keep it warm and direct."***

For a second, nothing happened. Then the grid rearranged like a sliding puzzle. Two focus blocks landed where Alex usually answered email. The appointment with Alex's boss moved to Friday with a short message that started, "Could we shift our check-in? I'm preserving a deep-work slot to hit the milestone you care most about..." Alex felt the day change by ten degrees.

"Thursday pickup's on me," Alex called back. And it was, because the hour that always went missing had been returned—without Alex staying up late to buy it back.

This is what the **Productivity Singularity** feels like when it's more than a buzzword. It isn't a science-fiction finish line, but a small hinge you can feel—a bend in the week. We arrive at it the way we arrive at most turning points: by noticing that the old math has stopped adding up.

For generations, *work harder* was the answer. Wake earlier, stay later, swallow the weekend and call it hustle. Our calendars turned into proof of virtue. Our inboxes became a running tally of unfinished promises. But history's true inflection points never begged us for more willpower. They bent the curve. The printing press didn't plead for better penmanship; it made copying irrelevant. The steam engine didn't ask for stronger backs; it multiplied force. The internet didn't ask us to memorize more; it connected minds at light speed.

We're standing at the edge of another bend. **Artificial intelligence isn't stealing your desk; it's redistributing effort**. Work that once required hours of attention—assembling, formatting, summarizing, restating—can now begin with a tireless helper that never gets bored. Your job is to direct it.

The Bend You Can Feel

A few blocks from Alex's kitchen table, there's a café where the afternoon light bleeds through the window like gold spilled across the floor. Elena sat at the corner table with a latte she kept forgetting to drink and a notebook decorated with swatches of color. She'd been inching her business forward for months—product photos at midnight, descriptions that never sounded quite right, email sequences she'd promise herself she'd finish on Sunday and never did.

Elena opened Midjourney and typed the prompt:

"Minimal, warm, nature-inspired packaging for small-batch skincare—neutral palette, soft texture, modern serif logotype."

The screen filled with four versions of the brand she saw in her head. Two didn't work. One was close. One felt like a future she could walk into. She exported and jumped to Canva, used the generative brand kit to carry the same feeling onto a label, a thank-you card, and a social header. Then she dropped a prompt into ChatGPT:

> ***"Write a 120-word product description in a friendly, confident tone for busy parents. Include three benefits and a one-sentence usage tip."***

It wasn't her voice yet, but it was a start—and starting had been the hardest part.

By the time the latte cooled, there was a storefront outline in Webflow, a first-pass copy she could edit in ten minutes, and a five-email welcome sequence drafted by HubSpot Breeze that began with value instead of a coupon. Elena didn't outsource meaning. She outsourced friction. And friction, as it turned out, had been most of the delay.

Across town, in a classroom that still smelled like dry-erase markers, Ms. Perez stared at three reading groups that needed three different entry points into the same story. Before the bell, she asked Khanmigo for a differentiated lesson plan that hit the standard without losing the kids who were ahead or the ones who needed one more rung on the ladder. She printed two sets of quiz questions: same concept, different scaffolds. When a student finished early, she asked for a fresh variant with one click. It wasn't magic. It was cycle time shrinking—minutes reclaimed and reinvested where they'd matter most: watchful eyes, well-timed encouragement, an extra beat of patience.

On the fifteenth floor of a glass building with air that always felt a little too cold, Omar pulled up a dashboard he'd been babysitting for years. He was the person who never missed a number, who sent perfect spreadsheets at 7:58 a.m., who tried to answer executives' tangled questions by Thursday even when they asked them to be delivered the following Monday. Reliable, invisible.

He turned on Copilot in Power BI and asked,

> ***"Show week-over-week churn by segment and highlight the top three drivers. Forecast the next eight weeks and give me a plain-English explanation."***

The chart redrew itself. The explanation came back in the kind of language that doesn't make people pretend they understand. Omar pasted the output into ChatGPT with one more instruction:

> ***"Turn this into a 150-word narrative for execs: what changed, why it matters, and what you recommend for the next sprint."***

The recommendation fit on one slide. For the first time, Omar arrived at the meeting with an answer and a path forward. Same person. New altitude.

And at a home office littered with colored pencils, Maria stared at a chapter her novel refused to give up. She'd never hand an AI model the keys to her voice; that felt like giving away the part that mattered. But she'd ask for a spark, prompting:

> ***"Give me five plausible motives for this character—two***

surprising, one morally ambiguous. One line each."

She didn't copy a single sentence. She underlined one motive and laughed at another and felt the old engine catch. The paragraphs that followed sounded like her. The pace returned. The story moved.

This is the Productivity Singularity when it's real: human intention at the center, machine capability at the edges, the curve bending where you need it most. Not less human. More.

From Curiosity to Stewardship

Most people arrive here the same way Alex did: curiosity, then a small experiment, then a result that's better than expected. The temptation is to stop there and call it a trick. But the real shift is what follows: stewardship.

Stewardship is the quiet skill of deciding **what gets your mind**. It's looking at the week and saying: these ten things are inputs and structure; these two are essence. You let the machine assemble, summarize, and shape. You keep judgment, story, ethics, and relationship—the parts that don't scale because they shouldn't.

There's a simple discipline that makes this stick:

- **Delegate** the first pass to AI when a task doesn't need your best brain (summarize, draft, cluster, propose).
- **Automate** what can run without you (move notes into

the right tools, schedule the recurring blocks, send the follow-ups).

- **Eliminate** what doesn't need to exist (the status report no one reads, the meeting that can be an email).

The mantra underneath is easy to remember when you're tempted to muscle through: **reduce before you produce**.

The Toolkit You'll Use Without Thinking About It

Stewardship becomes a habit when you have a few names for things:

- You run a quick **Friction Audit** every Friday: three places your time or energy leaked (inbox, meetings, handoffs, decisions). That's where you start.
- You set an **Automation Floor**—a baseline of tasks you won't do manually again (first drafts, summaries, formatting, transcription). The floor rises over time.
- You defend your **Human Ceiling**—the work only you can do (strategy, taste, empathy, ethics, relationships). You build walls around this time.
- You keep a **Review Ritual**—ten or twenty minutes a week to tighten prompts, adjust automations, and check that privacy and accuracy are intact.

Alex didn't call them those names at first. Alex just knew that after the calendar puzzle solved itself and the three reschedule messages were drafted and the Thursday pickup stayed in place, there was finally room for the one thing that'd been on the list for three months: finish the proposal that could change the team's quarter. The week bent upward.

Beyond Resistance: Working with Intelligence

There are reasons people hesitate to adopt this new way of thinking. One is the fear that if everyone uses the same tools, everything will look the same. That's only true if you ship the first draft. **AI is a rough stone. Your job is the cut and polish.** Use it to explore breadth and arrive at the second and third idea faster. Spend your specific taste on the final shape.

Other considerations to keep in mind:

- AI models can sound persuasive even when they're wrong. Treat them like clever interns—eager, tireless, and occasionally misguided. Verify high-stakes outputs against authoritative sources and qualified experts.
- Bias lives in data; don't let yesterday's patterns make tomorrow's decisions for people's lives.
- Privacy is borrowed trust; choose platforms with clear policies, turn on protections, and don't paste sensitive material into public systems without guardrails.
- Beware the tool treadmill. A thousand apps launch every quarter. The goal isn't to collect them. It's to pick the two or three friction points that hurt and fix those deeply before you try to fix anything else.

Ask yourself the right questions: **What deserves my best? What quality can only I add? Where will leverage help me be more human, not less?**

Elena's evenings returned and her conversion rate rose. That wasn't magic; it was cycle time and cost discipline compounding. If your cycle time is half a competitor's and you reinvest the time in more experiments, you don't just pull ahead—you pull away. If AI trims your administrative load by thirty percent and you channel that time into product quality and customer experience, you feel it in both revenue and morale. If you can prompt, automate, and synthesize across fuzzy problems, you move faster in the rooms where direction gets set. The gap that opens isn't *big versus small*; it's *adaptive versus static*.

Omar felt that gap when an executive asked, "What happens if churn rises two percent in our most profitable segment?" and, instead of promising an update "by end of week," he asked Copilot the question on the room's behalf and had a live answer before anyone refreshed their email. He didn't become someone else. He became the person he already was—but amplified.

There's one more piece no one can automate for you: identity. Many of us were rewarded for carrying the most—being the one who remembered, who could hold the plan in their head, who would stay late without being asked. That pride is earned. It's also a ceiling now.

The next decade rewards a different reflex: before I try harder, I'll reduce. Offload what a machine can hold. Keep for yourself the parts of the work that require nerve, discernment, and care.

Sarah, a project manager who'd perfected the art of being indispensable to everyone, finally admitted she was running out of hours she could steal from sleep. She gave Otter the job of catching decisions and owners so she didn't spend Sunday reverse-engineering action items from memory. She let Reclaim protect two deep-work windows like a bodyguard. She asked ChatGPT to turn bullet chaos into briefs that people would actually read. Ten hours returned in a week. She used five to do the work that'd been calling her name and gave five back to her life. Excellence without exhaustion didn't feel like a myth anymore. It felt like Tuesday.

If you want a glimpse of the near-term normal, picture small businesses run by a founder and a constellation of AI agents: marketing agents testing a thousand micro-experiments and keeping the dozen that convert; service agents resolving routine issues instantly and handing the rest to a human at the first sign of frustration; finance agents forecasting cash and nudging when reality drifts from plan; operations agents coordinating vendors without a status meeting. Inside larger organizations, **"Who are you?" quietly becomes "Who are you plus your AI?"** People show up with a personal stack tuned to their craft—prompts, micro-automations, a second brain of notes and patterns—the way earlier generations arrived with spreadsheets and playbooks. Thriving won't belong to people who memorize features. It'll belong to people who pair human edges—taste, empathy, ethics—with machine throughput to make something unmistakably theirs.

And if you're afraid of losing yourself along the way, hold the simplest anchor there is: leverage is dangerous only when it accelerates the wrong thing. Build your AI stack around your values—around the quality you want to ship, the relationships you want to protect, the health you want to sustain. AI won't erase you. It'll give you more time and more headroom to show up as yourself.

Back at Alex's kitchen table, the reschedule messages went out, the two focus blocks stayed in place, and the proposal that'd been carried from week to week finally moved from *someday* to *done*. When the Thursday pickup rolled around, Alex closed the laptop without the little punch of guilt that used to come with that click and said yes to the question that mattered. The week had bent. Alex had bent it.

Work harder carried us far. Work smarter carried us farther.

Work with intelligence—yours plus the machine's—will carry you beyond the old limits.

The engine is ready. The leverage is real. What you build with it is up to you.

2

Breaking Past Human Limits

You wake up to the same math you always have: twenty-four hours, one body, one brain. The clock hasn't moved in your favor. The calendar still fills itself with other people's priorities. Somewhere between the first notification and the last meeting, you feel the familiar constriction—like your day is a hallway with the walls inching inward. You've tried getting up earlier. You've tried the color-coded planner. You've tried the "just push harder" plan—twice. You know those tricks by heart. And you know their cost.

From Endurance to Leverage

For most of human history, people earned their edge by outlasting each other. They turned productivity into endurance sports—sleep less, stack more, sprint longer. The winners were the ones who didn't stop. That was the rule because it was the only rule we had. If you wanted more output, you paid in sweat and hours. You told yourself you were "disciplined," and sometimes you were. Mostly, you were tired.

Then something rare happened. The rule didn't exactly change; it widened. A set of tools arrived that didn't offer a twenty-fifth hour or a stronger body. They offered something stranger: more inside the hour, more from the body. Not magic. Not a rival person. A force multiplier for judgment. A quiet exoskeleton for ordinary work. A capacity amplifier for people who care about doing things well and still want to make it to dinner on time.

More Inside the Hour

If you want to see what this looks like, meet Sam. Mondays used to belong to spreadsheets, and everyone in the building knew it. Before lunch, Sam would migrate data across three systems, repairing broken Excel formulas like a plumber under a sink. After lunch, he became a slide factory—chart, title, callout, repeat. By five o'clock, he had a deck no one read carefully. He called it "doing the job."

A colleague asked him why Monday couldn't be a review instead of a full rebuild. They spent an hour one Friday connecting Power BI to the right sources and letting Copilot assemble the obvious visuals. Sam exported a simple CSV snapshot, dropped it into a prompt in ChatGPT—

"Summarize last week's performance, flag three anomalies, give me a 250-word brief with 'What changed / Why

it matters / What I recommend'"

—and went home skeptical.

Monday came, and the time Sam once spent wrangling data shrank dramatically. He still opened the dashboard, but now he started with the conclusion instead of the raw noise. Forty-five minutes later, he had a one-page memo that told a short, honest story. He didn't fill the reclaimed time with more reporting. He called three customers and asked the question the numbers couldn't answer: What did we miss? Two conversations later, he found the small operational change the dashboard had been pointing toward but couldn't say aloud. The work inside the hour compressed. His attention shifted toward value. He still did his job; he simply did the part that was his.

That's the texture of the change: less friction where friction teaches you nothing, more focus where you add value. Not speed for its own sake, but meaningful progress.

A physician named Lila would've told you she was fine, and you'd have believed her. She ran a clinic with quiet competence. She looked calm at the front desk and clinical in the exam room. But her day was a chain of small drains: rescheduling no-shows, sending patient-education materials, drafting follow-up notes, and answering the same six questions thirty different ways. None of it required her training; all of it siphoned the energy she needed for the parts that did.

She didn't "go full AI." She layered in a few practical changes: an appointment system that triaged rescheduling requests within set constraints; a library of approved patient messages that a model could personalize to each visit; and a dictation tool that turned her notes into clean documentation, proposed ICD codes, and produced a short patient summary while flagging uncertainty instead of inventing certainty. The effect wasn't a "robot doctor." It was a less depleted human at the end of the day—one with enough presence to be as careful with her last patient as with her first.

Energy Management

You can call this efficiency, but the real win is energy management. We pretend our minds don't tire. We plan days as if decisions have no weight and context switches don't extract a tax. They do. When you strip out the wasted effort that never improved your craft, you don't just go faster—you end with enough in the tank to do the piece you've been avoiding. You finally call the customer. You mentor the junior teammate instead of sending them a link. You walk outside without your brain buzzing.

Another kind of relief lives in attention. In the modern economy, **attention is the rarest currency we have**, and we spend it like someone else will top it off for us. We drown ourselves in dashboards, pings, previews, feeds. We confuse exposure with understanding. We tell ourselves we're "staying on top of it" while the real work quietly slides to the edge of the desk and waits for nine at night.

Marcos led a distributed product team and started each morning by triaging a hundred Slack messages, skimming five docs, and trying to reverse-engineer two inboxes into a plan. By ten o'clock he was tired, and the strategic thinking he was hired for drifted to after his kids were asleep. He didn't need a new personality; he needed a front door.

He opened Notion and built a home page for his head. The day's agenda lived on top. Priority emails were summarized by Superhuman Mail and routed into a single list. Jira's status updates fed a simple table, and three metrics—just three—that actually predicted whether the sprint was on track, sat in a small box at the right. He asked Notion AI each morning:

> ***"Give me the five decisions I need to make today. Include blockers, owners, and those three sprint metrics. Then suggest a ninety-minute deep-work block and defend it with my calendar."***

The inputs still existed; they just stopped shouting. Within a month, he felt something he hadn't felt in years: spaciousness. The work didn't get lighter. His mind did.

This is what leverage looks like in ordinary life. It turns a solo creator into a small studio. It makes a clinic feel like a coordinated team. It lets a teacher teach a room full of different learners without needing five copies of themselves.

Priya is a novelist who used to lose entire mornings to the blank page. She tried discipline; she tried all the rituals—tea, timer, playlist. On a Tuesday, she'd write one clean sentence and call it a victory. Then she tried ChatGPT for ideas with something smaller than "have a machine write the chapter," which she had no interest in doing. She asked for seeds instead of trees—conflicting desires between characters in a setting she loved, a handful of sensory details, three possible turns. Suddenly, the cold start was gone. She still wrote every sentence; she simply didn't have to dredge up the starting material alone. A year later, the short-story collection she'd been carrying in fragments finally existed outside her head. People complimented her "talent." She laughed, because it didn't feel like talent. It felt like momentum.

Rodolfo teaches a class of thirty-two students who read at five different levels. For years he dreamed of differentiating everything and lived with photocopied reality. He adopted an AI tutor and fed it the day's objective and reading text, then asked for five levels of scaffolding and questions that'd tell him who understood what. He didn't automate teaching. He automated whatever kept him from having his eyes on his students. The kids noticed. So did their parents.

The temptation is to conclude this is "doing more with less." That misses the psychology. T**he biggest drag on modern work isn't volume; it's micro-friction.** The five minutes here and six minutes there where nothing improves, but everything costs. When the scaffolding of life—calendar collisions resolved, travel stitched together, reminders arriving when action is cheap—stops living in your head, the head feels lighter. You still work, but you finally work where you matter.

Of course, this can all go sideways. If you point a fast tool at busywork, you'll get more busywork. If you speed up without standards, you'll ship out mediocrity in record time. If you fill reclaimed hours with more meetings, you won't feel any freer. The tool will do exactly what you ask with unnerving enthusiasm. **That means the gains hinge on intention.**

What's My Work, Really?

There's a question worth asking out loud: What's my work, really? Not the tasks you inherited, not the performative parts that look productive on a status update, but the value only you can add. For a leader, it's clarity, sequencing, and culture. For a clinician, empathy, diagnosis, consent. For a designer, taste and the courage to pick a direction. For a parent, presence. If a workflow or feature isn't buying you more time there, it's decor—impressive, maybe, but still decor.

Omar discovered this when he got lost in dashboards. He automated everything he could; models refreshed hourly; scenarios bloomed like wildflowers. He was producing more, but no one seemed to care more. The turning point wasn't a clever chart. It was a weekly note he started to send to his leadership team—one page, three sentences: what changed, why it matters, what he recommends. Executives stopped scanning pictures and started following a point of view. The machines did the math. Omar did the meaning. His job didn't vanish; it upshifted.

This is also where a quieter fear lives: If an assistant can do the part I find hard, does that part still matter? Calculators didn't make arithmetic irrelevant; they moved the bar for what counted as math. Grammarly didn't destroy writing; it raised the floor and put more weight on structure and voice. Copilots that draft emails won't turn you into a bland automaton unless you abdicate the part that's yours—tone, empathy, boundary. Use the lift to climb higher, not to stop climbing. The simplest guardrail is the most effective: **let the machine propose the first pass; you own the second.**

There are real traps beyond taste to consider:

- Over-delegation atrophies skill.
- Shiny tools fracture focus.
- "Clever" automations break at the worst time and steal an hour to save a minute.
- Data is borrowed trust; treat it like a friend's keys, not a toy.

The antidote isn't paranoia; it's practice. Small repetitions. Short experiments. Keep what helps; drop what doesn't. Treat curiosity as a habit, not a mood. If you don't like a feature, you simply don't have to use it.

Nadia's story is a checkpoint. A year into trying every tool she could find, she'd halved her hours on routine work and doubled her output. Her calendar looked like a neat garden. She was also more anxious than before. The extra time had become a vacuum she reflexively filled with more tasks—a treadmill disguised as progress. Her therapist asked the only question that mattered: **"What are you optimizing for?"** She didn't have an immediate answer, but she later wrote one out. The following quarter, her automation roadmap included a column she'd never considered: **What does this buy me?** If the payoff wasn't a clearer mind, a better relationship, deeper craft, or real rest, it didn't make the cut. The relief was both mental and physical. Her days didn't just get lighter; they got aligned.

Pull the camera back and you can see the horizon moving. Personal "second brains" remember more than we can. Tutors adapt faster than any one teacher. Creative models offer a hundred variations by the time you sip your coffee. Very soon, agents will act across services, and our days will be shaped by systems that anticipate rather than ask. The line between "my work" and "my tools' work" will blur, then blur again. When that happens, the assets that matter most aren't your macros. They're your standards, your stories, your relationships. Standards keep speed from eroding quality. Stories bring coherence to a flood of facts and convince human beings to move in a direction together. Relationships turn transactions into trust, especially when something breaks and a person—not a policy—has to step in.

None of this ignores risk. Bias hides in training data the way rot hides in wood; you have to test and audit before you lean your ladder against it. Tools speak in complete sentences, and we too easily mistake fluency for truth; skepticism is a feature, not a bug. Inequality won't fix itself; literacy and connectivity still gate who benefits. Those are decisions for leaders and citizens to make—in policy, in procurement, in classrooms. The existence of danger isn't an argument against power tools. It's an argument for using them with care.

So where does this leave you, today, at your desk or your kitchen table, staring at a list that outsizes the day? It leaves you with a new equation. Not "work harder," but work multiplied. Not "add hours," but raise the yield of each one. Not "be everywhere," but be fully here while your systems handle the scaffolding.

The steps are small on purpose. Choose one source of friction that quietly drains you. Introduce one tool that meaningfully reduces it. Route the saved time toward something you'll feel when your head hits the pillow. Repeat. Let compounding do what compounding does.

If you listen for the future of work, people will try to sell you the finish line: one-person corporations, agent swarms, neural interfaces. Maybe. But most of the life-changing gains will arrive disguised as ordinary days that feel less frantic and more focused. A meeting that ends with decisions because the transcript became action items before anyone asked. A night you close the laptop without guilt because the important work actually happened. A weekend that makes you a better weekday human. **The singularity won't announce itself with fanfare. It'll arrive as relief.**

We started with limits—time, energy, attention—and the honest admission that each has always capped what one person can do. We end with a different posture. Limits still exist; they just stop being the place where ambition goes to die. With the right help beside you, an hour isn't what it used to be, effort stops leaking through unnecessary seams, and attention can be guarded tightly enough to make work you're proud of.

If you can extend your limits, what will you extend them for? More output is easy. More meaning requires design. In the chapters ahead, we'll build those guardrails—how to keep your humanity intact as the tools get more powerful, how to steer speed toward health and relationships, how to translate compounding productivity into wealth you're actually glad to own. The tools will keep getting smarter. Your job isn't to keep up. It's to **point your newfound capacity at a life that feels like yours.**

3

Without Losing Your Humanity

You can trace the story of every major technology by the boundary it tests and the restraint it demands. The printing press broke the bottleneck on ideas—and let panic pamphlets race through towns. The telegraph collapsed distance—and made rumor outrun correction. The internet put libraries in our pockets—and pushed pop-ups through the same door. Now AI arrives with that same double edge, only sharper. It can accelerate learning, compress months into hours, and let one person move like ten. But it can also invite us, click by click, to stop using the very capacities that make us human—attention, judgment, empathy, imagination—until they thin from lack of use.

You feel that tension on an ordinary Tuesday. You ask a model to clean up your notes, and it does. You ask for a draft, and it hands you something polished. You aim a conversational agent at a customer queue and watch hold times drop like a stone.

"Just the boring parts," you tell yourself. It feels efficient. Helpful. Clean. And then one morning you open a blank document and hesitate. You're not blocked; you're distant. What slipped wasn't competence but contact—with your own thinking, with the people on the other end, with the why behind the work.

That's the quiet risk. Writing isn't only paragraphs; it's the struggle by which ideas become yours. Strategy isn't only tactics; it's the wrestling by which direction becomes conviction. Conversation isn't only information; it's the presence by which trust accumulates. **When we automate those struggles out of existence, we don't only gain time. We risk losing texture.**

The discipline that protects you isn't a tool but a posture. Use the machine as an exoskeleton for the mind—strength where you tire, not a replacement for the muscles you still need. Let it sift so you can respond with care. Let it draft so you can shape the argument. Let it map a field so you can decide where your field should go next. Keep your hands on the wheel where the road bends: ethics, stakes, relationships, taste.

You can see the difference most clearly in the places where the roles are right. In a radiology wing, a physician stands before a screen that highlights a cluster of pixels that'd have looked like fog yesterday. The model widens perception. But behind that screen sits a patient in a paper gown, joking just enough to steady themselves. The doctor walks in and does what a machine can't: names uncertainty without pretending it's certainty, sets a plan without stealing agency, and stays long enough for the patient to ask the real question. The AI model widened perception. The human reduced fear. The roles were right. Reverse them—ask a machine to console or a person to memorize millions of scans—and something essential breaks.

The same pattern shows up elsewhere. A junior developer rides the wave of a code assistant that scaffolds functions and threads imports before the coffee cools. It's sane leverage—until every solution starts somewhere outside their own head. Then a tricky bug lands, and the assistant hallucinates confidence with the wrong fix. The developer feels it immediately: the gap where a mental model should be. They switch the assistant off, trace the call stack by hand, and rebuild the shape of the system in their mind. The code works again. More importantly, the coder grows.

A writer faces the same truth at a desk. The model can produce fifty versions of the next paragraph, and each will be almost right—which is to say, wrong. The useful move isn't to surrender authorship but to use the tool as pressure against your own thinking. Ask for angles if you like. Ask for subheads to push against. Then write into the places the model can't reach: a smell from childhood only you remember, a confession that costs you, a sentence too strange for the average to attempt. The assistant can mimic genre and cadence. **It can't decide which risk is worth your name.**

Leaders face a subtler seduction. Dashboards whisper what customers feel. Productivity metrics whisper what people are doing. Slowly, you start to believe you can manage culture from a screen. Then one afternoon you catch yourself calling a listening tour a report. So you leave your office five minutes early for the one-on-one and wait in the hallway on purpose. Someone says the thing the survey never caught. You hear the tremor that tells you the team's brave face is brittle. The churn model was useful. The unguarded minute changed your mind.

There's an opposite mistake too. Stubborn purity can masquerade as virtue and collapse into fatigue. The craftsman who refuses every power tool wears out their wrists proving a point the work never asked them to prove. The scholar who prints a thousand pages to highlight by hand misses patterns a corpus analysis would've revealed before lunch. Tools exist to remove waste. Using them well isn't capitulation; it's stewardship. The real question isn't machine or human. It is: what's maintenance, and what's meaning? Automate the maintenance. Protect the meaning.

The Discipline of Staying Human

That protection works better when you set the line before convenience arrives to smudge it. You make yourself a small reminder card: I write student recommendations from scratch. I accept AI help with brainstorming, but our company's manifesto is human down to the comma. I never send an apology, condolence, or thank-you that a model wrote—those words will cost me the time they deserve. Ethicists call this pre-commitment. You can call it deciding in the morning. Boundaries set at 8 a.m. tend to hold better than heroics attempted at 5 p.m.

Then you build habits that keep your capacities fit. Attention is still the battery every technology depends on. Maya—manager, parent, human—used to bounce between inbox and chat until bedtime, then fall asleep with her jaw clenched. After automation, but before discipline, she felt different but not better: summaries on top of summaries, a calendar tiled like a bathroom wall, a brain that looked efficient and felt like a strip mall.

What helped wasn't another app. It was a sequence. In the morning, an assistant condensed Maya's messages into a single page: three decisions only she could make and five quick replies she could edit in her own voice. She read it once, closed every communication window, and moved into a ninety-minute block of human work—thinking, writing, deciding. At lunch, she walked without headphones and let her mind settle. In the afternoon, she used transcripts to extract actions, then sent three handwritten notes: two to colleagues naming specific moments they'd handled well, and one to a customer who'd been patient. Technology handled throughput. Human gestures built trust. Productivity rose. The team's anxiety eased.

At home, the pattern looks different but means the same thing. Luis and his partner stare at a week that doesn't fit: two kids in sports, one parent on shifts, and a budget allergic to takeout. An AI meal planner and calendar agent turn chaos into an itinerary—groceries ordered, prep steps spread across days, carpools reduced. But Friday dinner stays device-free and slow on purpose. The assistant can nudge. It can't notice the sulking teenager who needs to be drawn into conversation, or decide that a bad week deserves a good ending and a trip for ice cream. **Tools remove friction so human ritual can deepen.**

In a classroom, Reva teaches five reading levels in one hour. An AI tutor creates differentiated practice sets while she spends her time in the aisles, kneeling next to the student who's one failure away from quitting. "What makes you think that?" she asks. And the student hears the thing every AI model tries to simulate but never feels: you matter enough for someone to spend their time on you.

Across town, Jonah uses a generative model to storyboard shots and explore color. He's honest with his crew about what the machine rendered and what the camera will find. The time saved on palettes becomes time walking the city, asking strangers for stories that—with permission—enter the film's emotional DNA. When the model suggests a line reading, he ignores it on purpose. Performances have to be lived, not pasted. That's the point: modern, not synthetic.

Even with good intentions, though, another motive can creep in. Over-automation doesn't only promise efficiency. It promises relief—relief from uncertainty, from the discomfort of not knowing, from the vulnerability of being seen. Let the model write the apology and you don't have to sit inside your mistake. Let a recommender pick the strategy and you don't have to risk being wrong. Let a bot answer the phone and you don't have to feel another person's frustration. But growth tends to live on the other side of tolerable friction. Outsource every edge and, sooner or later, you outsource your own self-development.

So you keep an edge alive on purpose. You write one paragraph each day that no assistant touched. You whiteboard a problem by hand even when a plugin could draw it prettier. You sit with someone who's upset and resist the urge to fix too quickly. You listen until the story empties and ask the generous question: What would help?

This scales beyond individuals. Teams that thrive make their augmentation doctrine explicit. In writing, not vibes. They name where models may propose and where humans must decide. They define when to escalate from chat to call to in-person. They set ways to catch and remediate model errors, protect privacy, and guard the work that's core to identity. They train people not only which buttons to click but how to think about technology so trust is earned and kept. And they measure more than throughput: burnout, trust, recovery after errors, the things that determine whether speed is sustainable or corrosive. Companies that deliver at machine speed while depleting humans can look brilliant for a quarter. Then they quietly burn out.

The Future Worth Choosing

There's a civic layer, too. The same nudge that helps one person get to bed can push someone else into a rabbit hole. The same routing that speeds packages can shape policing. The same avatar that comforts a lonely person can be misused to extract sensitive information from a child. Guardrails can't live only at the level of personal ethics; they also have to become public standards. We need disclosure when AI influences consequential recommendations. We need audit trails in housing, hiring, credit, and parole. We need privacy standards that don't hide consent inside a thicket of legalese. We need to raise the standard from "it works" to "it works and preserves dignity." That isn't anti-innovation. It's what allows innovation to last.

And if that sounds heavy, remember the upside. When you let machines lift what they're built to lift, you get your hands back for the work only you can do. Kevin, who runs support at a growing brand, replaces a brittle chatbot with a hybrid system: AI for lookups and order status, a handoff at the first sign of confusion or anger, and a small budget for human gestures—proactive refunds, thank-you notes tucked into replacement orders. Ticket time drops. So does turnover. Customers feel seen. Reps feel permitted to care. The point becomes obvious: the tool exists to make generosity easier, not to make people disappear.

Amira, a solo creator, is drowning in the cadence she thinks the algorithm demands—daily posts, relentless replies, constant engagement. She uses scheduling and a writing assistant to batch the routine, then spends the recovered time on what can't be faked: unscripted live Q&As, voice notes to new customers, handwritten cards. Growth stalls for a month. Then the community deepens and referrals start arriving in plain daylight. The signal isn't volume. It's intimacy. **AI made it possible not to perform more, but to be more present.**

Omar, the analyst who automated his spreadsheet grind, doesn't become the person who emails prettier dashboards. He becomes the person who notices unexpected relationships and asks better questions. His promotion comes not because his charts look cleaner, but because his thinking does. The tool freed him to climb from tasks to insight.

A simple test settles into your pocket. When a new capability arrives, ask: Will this help me be more present where presence matters? Will it protect my attention so I can spend it on judgment and creation? Will it preserve or enhance the dignity of the people it touches? Will it make me more capable over time, not less? If the answers lean yes, adopt boldly. If not, restrain yourself—even when the demo dazzles.

You won't get that balance right every week. You'll send a note that sounds like a machine and feel a flicker of embarrassment when the reply assumes a warmth you didn't actually bring. You'll delegate a conversation that needed your face. Forgive yourself quickly. Recalibrate promptly. Humanity isn't a trophy you win once; it's a bias you practice through a thousand small choices.

And when we get it right, the reward isn't just speed. It's truth. Automation becomes a form of care—for your future focus, for the quality of attention you bring to people, for the craft you refuse to outsource. The things we fear will vanish—originality, empathy, grit—don't disappear. They concentrate. The darker alternative isn't hard to imagine: efficient emptiness. Work that ships faster and says less. Teams that execute crisply and trust less. Leaders who hit the numbers and still can't sleep. A society that optimizes its way into loneliness. That future isn't inevitable. It's a collection of choices we're still free to reject.

So let the machine be a partner. Invite it into the parts that strain your wrists and your patience. Ask it to shorten the path from a flood of data to a clear picture, from logistics to ritual, from first draft to thoughtful second pass. Then, inside the space you win back, do the slow work: write the paragraph that costs you something because it's true; make the call that creates safety because it's brave; stay with the problem until its shape changes because you did. **The productivity worth chasing isn't one where humans vanish into their tools. It's one where hu-**

mans finally have enough time and headroom to be fully human.

Automate what drains you. Guard what defines you. Use speed to reach wisdom sooner. Use scale to reach people better. Keep your hands on the wheel where the road matters most.

Part II – Productivity Reimagined

"Should you find yourself in a chronically leaking boat, energy devoted to changing vessels is likely to be more productive than energy devoted to patching leaks."

– Warren Buffett

4

Automating the Mundane

You don't notice the first drop. You open your laptop and there it is—one more "quick" reply, a reschedule, a fresh thread that looks like it needs your eyes. You copy a figure from one sheet to another. You promise yourself you'll organize those notes later. You approve a calendar hold because finding a time feels harder than losing one. None of it is dramatic. None of it feels like a decision. But by 11 a.m., you're drowning in drip-work, and by late afternoon, you're soaked. The work that'd actually move the needle is still at the edge of your desk, waiting for the weather to clear.

We tell ourselves we can out-hustle the drizzle. We plan to get up earlier, grit our teeth through the inbox, and "catch up tonight." Then tonight arrives, and we have no energy left to catch up. The loss isn't only time; it's cognitive. Every micro-interruption taxes your attention. Each context switch scrapes away a little focus.

You start with a plan and end with a scatterplot. You weren't lazy; you were rained on.

For the first time, you don't have to out-grind the weather. You can route around it.

The change starts small. Email stops being a museum of newsletters when an AI assistant distills the top 20 into five decisions with drafts you can approve in seconds. Scheduling stops peeling hours off your week when a calendar agent negotiates within your constraints and defends the 90-minute blocks where your real work happens. Slides you rebuilt every Thursday become a live dashboard tied to source data, quietly updating while you sleep. The meeting that used to evaporate into folklore leaves behind a decision log and three owners, posted to the place your team actually looks. The day doesn't get louder; it gets clearer.

Sarah felt the first clear morning like that in week three. She managed cross-functional projects at a global firm and loved solving problems. She hated coordinating ten calendars and rewriting meeting notes for people who missed them. On a Friday, she wrote the simplest plan she could live with: guard one deep-work block each morning, let Motion and Reclaim negotiate the rest; funnel updates into one Notion board; turn on Otter in every meeting and push action items to the task list people already used. She added one constraint: weekly stakeholder summaries had to fit on a single page.

At first it felt mechanical. By the second Monday, she noticed the shape of her day changing. The 9 a.m. block stayed quiet.

The 10:45 handoff landed in the right queue without a DM. By the end of the month, she had a whole workday back every week. Delivery improved, sure, but the change that mattered showed up in places dashboards can't measure. Evenings with the laptop closed. Weekends without "just an hour." Her ideas sharpened because her attention wasn't frayed. Her tone softened because she wasn't at the edge of her nervous system. Her team noticed before she did.

David's weather was different. He built a solo business on grit and long Saturdays. He prided himself on personal replies and thoughtful content, but admin smothered both. He didn't want to hand his voice to a bot, so he started gently. A Zapier flow routed the most common support questions to AI-drafted replies he approved in batches. Jasper roughed in blog posts and sales emails; he tightened them until they sounded like him. Help Scout surfaced only the tickets that truly needed his judgment. The first weekend he tried it, nine hours of reactive grind became ninety minutes of high-leverage editing and an afternoon mapping partnerships he'd postponed for a year. Revenue didn't spike overnight. Anxiety dropped quickly. With space to think, he chose better projects and priced with less fear. Growth followed the relief.

Priya loved the human side of HR and felt trapped in paperwork. AI didn't replace her taste or ethics; it gave them room. A screening layer sorted résumés against actual criteria instead of keyword noise. Drafted job posts reflected inclusive language and clearer expectations. Onboarding packets assembled themselves from templates. Candidates noticed faster, kinder communication. Inside the company, Priya finally had margin to sit with managers, run listening sessions, and coach hard conversations—the work that prevents regrettable exits. Her influence didn't shrink; it expanded, because she had time for the human parts only she could do.

Across all three, the pattern repeats. **AI didn't make anyone lazy. It made them leveraged.**

The Leverage Loop

The leverage lives in a simple loop you can run all day without drama: delegate, automate, eliminate.

You're staring at a 50-message email thread. Delegate the first pass. Ask an AI model to condense it into three decisions with pros, cons, and a draft reply for each. You make the call. Your name stays a guarantee of quality. An hour later, you're looking at a form submission that'll trigger three handoffs. Automate the path. A Make scenario or Zapier Zap can create the task, assign the owner, post the update, and close the loop without you forwarding anything. That afternoon, a recurring status deck wants to be built because it's always been built. Eliminate it. Replace slides with a live dashboard and a one-page weekly brief. If no decision depends on a ritual, kill the ritual. Your week will exhale.

This isn't about mastering shiny tools. It's about reducing maintenance so you can return to meaning. Your brain can hold an arc again. You stop thinking in fragments. Strategy improves because working memory isn't packed with lint. Writing gets cleaner because it isn't interrupted every eight minutes. Leadership gets kinder because you aren't stretched thin enough to snap. Health steadies because sleep isn't the only time your laptop is closed. You didn't become someone else. You finally created conditions where you can be the person you meant to be at work.

Zoom out from your day to your system and the leverage compounds.

Most teams don't lack intelligence; they lack shared memory. Ideas live in inboxes. Decisions live in heads. Work lives in private documents. Every question starts a scavenger hunt. You can't change culture with a plug-in, but you can make the right behavior easy. Put sources where they belong—one Notion or Airtable base as the default, not the exception. Let AI summarize discussions into decisions and push them to the place everyone will see them. Train a lightweight internal copilot on approved internal artifacts so a new hire can ask, "What's our stance on X? Where's the latest Y?" and get an answer with links instead of folklore. The first time a teammate finds an answer without sending another "Got a sec?" message, you can feel **compound clarity** beginning.

Clearing the Weather

There are pitfalls, and they're predictable. Over-delegation turns you into the person who forwards AI output without reading it. People notice. Keep your name a promise. Tool-chasing fractures attention and prevents mastery. Pick a small stack and go deep. Privacy laziness—copying sensitive data into whichever free tool is nearest—will cost you trust. Use an approved, appropriately secured platform. Read the policy. Treat data like borrowed dignity. And beware invisible toil disguised as "quick checks." If it steals an hour a week, it deserves a proper fix. If it "just takes a minute," that minute is probably multiplied across ten people.

There's a psychological tension worth naming. Busyness is an identity that pays in applause. Many of us learned to measure our value by how much we carry. Offloading the mundane middle can feel like stepping off a treadmill while everyone else keeps running. It can feel like cheating. It isn't. It's leadership. You aren't paid to route low-value work. You're responsible for outcomes and for the health of the system that produces them. If you can achieve the same outcomes with less human strain, you've done your job well.

You'll feel the shift before anyone congratulates you for it. The day will get quiet in places that used to buzz. Meetings will end with decisions because the transcript turned into actions before anyone asked. Your calendar will defend what matters without requiring a daily fight. You'll write fewer emails and say more with them. Your team will breathe. And in the quiet, you'll remember what you're building.

One afternoon, a new hire asks where to find the latest metrics. You don't answer with a paragraph; you paste a link to the single source of truth and the internal copilot thread that explains each definition. "Thanks," they say, and move forward without another scavenger hunt. That's the sound of weather clearing.

The point was never to do more for its own sake. It was to

do less of what doesn't matter so you have more energy for what does. It was to shift the weight from maintenance to meaning. And once the mundane middle is out of your way, the most useful question reappears—the one busyness conveniently kept at bay:

With space to think again, what will you choose to build?

5

Decision-Making in the AI Era

For a long time, the hardest part of making a decision was getting your hands on the pieces. The numbers were somewhere else. The notes lived with another team. The customer feedback sat scattered across support threads, survey comments, screenshots, and someone's half-finished summary in a shared drive. By the time everyone finally had enough of the picture to talk about it, the meeting was already tired. The clock was low. The energy had thinned. Sometimes the decision got made anyway, but often it didn't. It just drifted into the polite language of delay: let's revisit next week, let's get a little more data, let's socialize this first.

That used to feel normal because information itself was expensive. Gathering it took time. Organizing it took more. Carrying it all in your head felt like leadership.

But that isn't the world you're in anymore.

Now the problem isn't always that the pieces are hard to find. More often, the pieces arrive too fast. A tool can summarize a stack of documents before the meeting begins. It can read through customer comments, cluster themes, surface objections, draft options, and sketch the risks in a few seconds. That changes the bottleneck. The work is no longer only about access. It's about discernment. It's about knowing what kind of decision is sitting in front of you, what deserves speed, what deserves caution, what's reversible, what isn't, and what values are quietly hiding underneath what first looked like a data question.

You can feel this shift in an ordinary room.

A team gathers around a product question that should be straightforward and somehow isn't. Revenue is involved. Customer trust is involved. Engineering capacity is involved. The first ten minutes disappear into context. One person is talking about roadmap timing. Another is talking about support volume. A third is pulling a dashboard that half the room can't read on the fly. It'd be easy, in a moment like that, to assume the answer is one more meeting or one more analysis. But often the real problem is simpler than that: the room hasn't yet named the decision clearly enough to think about it well.

That's where AI becomes useful in a way that's more interesting than "answering questions." It can help turn a vague, overloaded situation into something with edges.

Someone opens ChatGPT and types:

> ***What's the real decision here? What are we mixing together that should be separated? Give me the three main trade-offs in plain English.***

In a few moments, the room is looking at a problem with cleaner lines. What felt like one giant decision turns out to be three smaller ones. One is about timing. One is about risk. One is about values. The confusion doesn't vanish, but it stops being shapeless. And that matters, because people can reason better when the decision has bones.

That's one of the quiet revolutions of decision-making in the AI era. The best use of the tools isn't to replace judgment. It's to help judgment arrive sooner, less exhausted, and standing on firmer ground. The machine gives you a first cut of order. The human still has to decide what deserves belief, what deserves skepticism, and what deserves care.

The temptation, of course, is to treat speed as wisdom. The faster the summary appears, the more authoritative it can feel. But a fast answer isn't the same thing as a sound answer. If anything, the ease of generating plausible language makes the human job more serious. Someone still has to ask whether the recommendation is real or merely polished. Someone still has to ask what was ignored, what assumption is doing too much work, what cost has been made invisible because it's easier not to name it.

That's why the role of the decision-maker changes. The strongest person in the room is no longer necessarily the one who can carry the most information. It's the one who can hold the room steady long enough to ask better questions. What kind of decision is this? What would change our mind? What would failure look like early? Who carries the downside if we're wrong? Is this actually a data problem, or is it a values problem disguised as analysis?

Once you begin asking those questions, the atmosphere changes. What used to feel like delay starts to look like avoidance. What used to look like decisiveness sometimes reveals itself as impatience in a suit. And in the space between those two mistakes, a better rhythm begins to form.

The Decision Starts to Breathe

Lila, a product leader, feels this shift most clearly on a Tuesday afternoon when a major customer request collides with a reliability problem no one can keep pretending is minor. Sales wants the feature. Engineering wants time. Support wants relief. The room is full of valid concerns and already tilting toward a familiar outcome: everyone makes a decent case, no one wants to own the wrong call, and the meeting stretches into the swamp between urgency and caution.

Instead of letting the conversation harden into roles, Lila opens a blank page and gives the room something firmer to stand on. She asks ChatGPT to summarize the last month of support tickets, customer complaints, and incident notes. Not everything—just the parts that matter most. What's getting worse? Which customers feel it most? What changed recently? What problem are people actually describing when they say the product feels broken?

What comes back isn't a verdict. It's something more useful: a shape. Complaints about slowness are rising. Larger customers mention it more than smaller ones. The pattern sharpens after a recent infrastructure change. The team is still facing a trade-off, but now it's facing the same trade-off together. That's no small thing. A surprising number of slow decisions happen because the room never quite agrees on what reality is.

Then Lila pushes further. She asks for three credible paths forward: ship the feature now, stop and run a reliability sprint, or split the difference and do both more slowly. She asks for the likely upside and downside of each, not because the tool can know the future, but because surfacing the logic helps the humans test it. Then she asks the question that matters most: If we choose reliability first and regret it, what will probably be the reason, and what would we notice soon enough to change course?

That question changes the room. Suddenly the decision doesn't feel like a referendum on courage or caution. It feels like a move with conditions. If the reliability work improves the system fast enough, the team earns the right to ship into calmer water. If it doesn't, they pivot. The emotional burden of the decision lightens, not because the stakes disappear, but because the team has turned a foggy commitment into something closer to an informed experiment.

That same pattern repeats elsewhere, in places that look nothing alike on the surface.

Victor, an operations lead, sees it during an incident call when payment errors spike and ten people nearly start guessing at once. In another era, the call would've turned into noise fast—half-tried fixes, overlapping theories, someone insisting on intuition because intuition feels faster than method when the graph is turning red. Instead, Victor pastes sanitized logs and monitoring snapshots into an approved instance of Claude and asks a calmer question: What are the most plausible causes, and what's the safest way to test each one? The answer isn't magic. It doesn't solve the outage. What it does is restore sequence. The team stops thrashing. One test leads to the next. The noise drops enough for signal to emerge. By the time the system stabilizes, the room is no longer merely relieved. It's clearer. They know what happened. They know why. They know what to watch next time. The machine didn't own the incident. It shortened the distance between panic and useful action.

In a clinic across town, the same logic takes a gentler form. A physician-administrator team is debating whether some routine post-operative visits could move to telehealth. The surface case is obvious enough: less travel for patients, easier scheduling, and a lighter burden on the practice. But everyone in the room knows the real question isn't efficiency. It's whether convenience can be gained without sacrificing dignity, reassurance, or safety. Before anyone reaches for a conclusion, they use Perplexity to identify relevant evidence on outcomes, satisfaction, and risk, then verify that evidence against authoritative medical sources. Afterward, they do something AI can't do for them: write their values at the top of the page in their own hand—equity, continuity, informed consent.

That act matters more than it sounds. Once the values are visible, the conversation changes. The decision is no longer, *Can we do this?* It becomes, *How would we do this in a way that still feels honest to who we're responsible for?* The answer isn't a sweeping rollout. It's a pilot. A narrower move. A version with guardrails. Again, **AI doesn't decide.** It helps compress the evidence and the options so the human beings in the room can spend more of their energy on the part only they can do: weighing the trade-offs in a way they're willing to live with.

The same thing can happen in a quieter setting, at a kitchen table.

A family is trying to decide whether to move their child to a different school. On paper, the new option offers stronger programs and shorter waitlists. The current school offers familiarity, friendships, and teachers who already understand the child. Left alone, the conversation could easily swing between hope and guilt, ambition and protectiveness, the future and the tenderness of what's already known. Instead of letting the decision stay vague and emotional in all directions at once, they gather the plain facts first—commute time, class size, cost, support services—and use ChatGPT to organize them into something simple enough to look at without spiraling. Then they move toward the more human facts: what the child is excited about, what they fear, what a normal Tuesday might feel like in each place. They write down what they're really trying to protect: safety, curiosity, friendships, budget, time together.

They still do the real-world work. A shadow day. A real commute. A conversation with another family. But now there's a clarifying question on the table that helps them think in a straighter line:

If we switch and regret it, what will probably be the reason—and what would we notice first?

The decision is still hard. It should be. But it stops being a perfect answer chased into paralysis and becomes something better: a **good decision with a review date**.

This is what AI enables when it's used well. Not certainty. Not wisdom by outsourcing. Not relief from the burden of choosing. What it offers is a way to move from swirl to structure faster, so the actual human work of deciding can happen with more honesty and less drag.

What the Tools Can't Carry

The more capable the tools become, the easier it's to forget what they can't do.

They can't care.

They can't absorb the moral weight of a decision. They can't tell you how much trust is worth protecting, how much strain is too much to ask of a team, how much family time you're willing to trade for more money, or whether the efficient answer is also the decent one. They can produce language around those questions. They can make the trade-offs easier to see. But they can't own the consequences. Only people can do that.

That's why judgment becomes more visible, not less, in the era of AI.

When information gets cheap, what matters isn't merely access to it, but taste in what to trust. When options multiply, what matters isn't how many paths you can generate, but whether you can tell which ones are real. When summaries arrive instantly, the danger isn't just laziness. It's the quiet illusion that because something has been compressed, it's been understood.

You can see the teams that handle this well because their conversations start to sound different. They're less impressed by polished confidence. They ask what would change the recommendation. They ask who benefits and who absorbs the risk. They notice when a values question is trying to hide inside a data costume. They know that some choices are reversible and should move quickly, while others deserve a slower hand because dignity, trust, or identity sit in the room with them whether or not anyone says so.

Over time, that discipline becomes cultural.

The meeting begins with a clean brief instead of a cloud. There are a few real options, not ten decorative ones. Someone names the trade-off plainly. Someone marks what will be watched after the decision, not just before it. A review date goes on the calendar. The note afterward is simple and human: this is what we chose, this is why, this is what we're watching, this is when we'll revisit.

None of that's flashy. That's part of why it works.

The work stops vibrating so much. People breathe more easily. There's less ritualized confusion and less worship of delay simply because delay sounds mature. There's also less false bravado, because the room is no longer forced to choose between hand-wringing and swagger. It has a middle path now: clarity, testing, commitment, review.

And in that quieter, cleaner rhythm, another question eventually appears.

Once the tools have made it easier to gather the evidence, clarify the options, and test the assumptions, what remains isn't a technical question at all. It's a human one.

What are you trying to become better at choosing—and what should those choices serve?

A company can use AI to accelerate bad instincts just as easily as good ones. A leader can use it to dress up avoidance in better language, or to make room for actual courage. A family can use it to organize a difficult choice, but still has to decide what kind of life they want that choice to serve.

That's the heart of decision-making in the era of AI. The clutter gets cheaper. The options come faster. The cost of getting oriented drops. And because of that, the human job becomes easier to see and harder to avoid.

You still have to name the real choice. You still have to test it honestly. You still have to move with your eyes open.

The tools can help you do that sooner.

They can't do it for you.

6

Reprogramming the Mindset for Exponential Growth

On Monday morning, someone can open ChatGPT, Claude, or Copilot and do in ten minutes what used to take half a day. They can compress a fifty-page report into five decisions. They can turn a messy set of notes into a clean first draft. They can ask for three plausible ways through a problem that's been stuck for weeks. They can translate, summarize, brainstorm, red-team, and reorganize before the coffee cools.

And then, by Friday, they're back where they started—juggling windows, apologizing for delays, and promising themselves they'll "catch up tonight."

The tools didn't fail. The operating system between their ears didn't update.

That's the real story of this era. The leap isn't only technical. It's mental. AI doesn't change a life just because it exists on your laptop. A person can carry astonishing capability in their pocket and still live exactly as they did before—reactive, overextended, loyal to habits that made sense in a slower world. The bottleneck is no longer only access to intelligence. It's the mindset that decides whether intelligence becomes leverage or just one more tab.

We were trained for a world that moved in straight lines. Effort in, result out. Years of repetition, then mastery. Step one, then step two, then step three. That model made sense because progress used to feel more visible and more proportional. You worked longer, practiced harder, waited your turn, and eventually the line advanced.

But the line doesn't behave that way anymore. It bends. It bends in technology, in business, in media, in learning, in the speed with which good ideas scale and mediocre ones disappear. What felt advanced five years ago now feels ordinary. What feels advanced now may feel quaint before long. AI doesn't create that curve by itself, but it makes it impossible to ignore. The slope is steeper now.

And that's where many people get trapped. **They keep applying a linear mindset to an exponential environment**. They wait to feel fully ready. They treat each new capability as a curiosity instead of a signal. They assume change will arrive gradually enough to negotiate with. So the future lands not as an opportunity but as a sequence of **ambushes**. One day a task takes half the time it used to. The next, the baseline for what counts as competent has quietly moved. The day after that, someone else is doing with one person and three AI systems what used to require a full team and a week.

The invitation isn't to panic. It's to reprogram your expectations.

Instead of asking, *Do I like this tool?* ask, *What happens if this capability improves again, and again, and again?* Instead of asking, *Can I ignore this for a while?* ask, *Where do I want to be standing if this becomes normal faster than I expect?* That's the mindset shift. You stop evaluating AI as a gadget and start seeing it as part of the environment—like electricity, like the browser, like the phone once was. Something that changes the shape of what's possible, and therefore changes the shape of what your habits must become.

That sounds abstract until it lands in the middle of an ordinary week.

A manager sits down Monday morning to prepare the same weekly update she's prepared for years. The ritual is familiar: open three spreadsheets, chase two missing numbers, rebuild the slide that always breaks, then write a summary no one reads until the meeting is already underway. In the old mindset, the question is how to push through the ritual faster. In the new mindset, the question is whether the ritual deserves to exist in its current form at all.

So she tries something small. She connects the source data to a live dashboard. She uses Copilot to draft the first summary of what changed, why it matters, and what needs attention. She asks ChatGPT to turn a wall of KPI movement into a short brief with risks and suggested actions. It isn't elegant on the first pass. A field name is wrong. One insight is obvious. Another is too polished to trust. But the point isn't perfection. The point is that, for the first time, she's no longer spending all her energy on assembly. She's bought back the ability to think.

That's the first taste of exponential leverage. Not magic. **Recovered judgment.**

The Small Loop That Changes Everything

Most people imagine a mindset shift as a dramatic revelation. It rarely is. More often, it's a sequence of small experiments that teach your nervous system a new truth.

The writer who believed help would dilute her voice discovers that ChatGPT can be used not as a ghostwriter, but as a spark plug. She asks for ten possible character tensions, rejects half of them, keeps two, rewrites everything, and realizes the work still sounds like her because she's still the one carrying the risk. The model didn't replace her voice. It got her to the edge of the page faster so she could spend her energy on the part only she can do: **cadence, image, nerve**.

The coach who insisted he worked with people, not systems, eventually asks a humbler question: what's stealing the hour I most need? Not in theory. This week. The answer isn't "coaching." It's the swarm around coaching—scheduling, check-ins, workout logs, missed follow-ups, the low-grade administrative moss that grows over any human business if no one clears it. So he lets Motion protect deep-work blocks, uses automation to route routine reminders, and asks Claude to notice patterns in client adherence logs before he does. The machine doesn't tell him what matters. It points. He cares better because the clutter thinned.

The teacher facing a room with ten different levels of readiness doesn't need grand ideology. She needs help building three versions of today's lesson before the bell rings. So she asks ChatGPT for challenge problems, scaffolded examples, and misconception checks. The output is useful, awkward, incomplete. She revises the tone. She adds the analogy that fits her students. She folds in the question she always asks when she wants them to think instead of perform. What changes isn't simply the worksheet. What changes is that she's no longer spending all her energy manufacturing the middle. She has more left for the human work: watching faces, adjusting pace, catching the quiet kid before they disappear.

That's how the rewiring happens.

Not through one giant leap, but through a loop so small it almost hides inside the day:

Notice. Test. Reflect.

You notice the old reflex. *This is hype. This isn't for me. This is cheating. This is too much. We tried something like this once.* That voice isn't evil. It's your system protecting the familiar. But the familiar isn't always the same thing as the wise.

Then you test. Not with a transformation plan. With fifteen minutes. Draft one email with ChatGPT and compare it to the way you'd have done it alone. Let Reclaim propose times for a week and see if the sky falls. Ask Claude to argue against the decision you think you've already made. Use Otter to catch a meeting's actions so you can see what it feels like to end a conversation with three owners and three dates instead of good intentions and fog.

Then you reflect. Not vaguely. Specifically. Did this save time? Did it reduce friction? Did it clarify something that'd otherwise have remained muddy? Did it make the next decision easier, the next meeting cleaner, the next hour more human?

If the answer is yes, you don't merely admire the result. You capture it. Save the prompt. Standardize the workflow. Put the behavior on the calendar. Turn the experiment into a repeatable pattern. That's when a tool stops being a novelty and starts becoming part of your **automation floor**—the layer of work the system can hold so that you can spend more time being human: judgment, relationships, creativity, decisions.

This is where exponential growth becomes less mystical and more lived. **It isn't just that each tool saves a few minutes; it's that those minutes stack into better focus, which improves decisions and outcomes.** Better outcomes reduce panic, and less panic improves judgment. Suddenly, the growth isn't only in output. It's in posture. You're less frantic, less linear, and less dependent on brute force.

That's why the deepest trap isn't being bad at tools. It's thinking the point of AI is to do more of what you're already doing.

It isn't.

The point is to ask what should be eliminated, what should be automated, what should be amplified, and what should remain deliberately, stubbornly human. **The person who keeps the same life and simply adds AI on top will feel busier in a more modern way. The person who uses AI to rethink the shape of life itself begins to move differently.**

Becoming Exponential Without Becoming Hollow

There's a fear underneath all of this, and it deserves more respect than people usually give it.

The fear isn't only that the machines will get better. It's that in the process, you'll get blurrier. That your voice will flatten into polished sameness. That your value will shrink. That everyone else will adapt faster and you'll always feel one cycle behind. That the tools won't just change how you work, but quietly erase what makes the work feel like yours.

That fear can't be argued away with optimism. It has to be answered with contact.

You write with ChatGPT and discover the draft is stronger but still clearly yours once you put your hands back on it. You let Motion or Reclaim clean up your scheduling and notice that relationships improve because you're less scattered, not less human. You ask Claude to red-team a plan and realize the machine isn't replacing your judgment; it's strengthening it by surfacing the blind spot before it becomes expensive. You use Cop**ilot** to give you the first pass on a briefing memo and realize the real value isn't that it wrote faster. It's that you had enough energy left to think harder.

Little by little, the fear changes shape. It stops being, *Will this erase me?* and becomes, *How do I use this to become more fully myself?*

That's the right question.

Because direction matters more than velocity. AI can wrap your days in cleverness and still leave you hollow if all you're multiplying is noise. A faster version of drift is still drift. A more automated version of the wrong life is still the wrong life. Beneath all the tactics sits a simpler question: Why are you multiplying?

If the answer is simply more, faster, bigger, the curve will own you. You'll optimize yourself into exhaustion with better software. But if your aim is to do work you're proud of, serve people better, make cleaner decisions, and buy back time for what matters, the curve begins to serve you. It becomes a source of leverage instead of panic.

That's why the mindset upgrade matters more than the individual app. The names will change. Today it's ChatGPT, Claude, Copilot, Otter, Motion, Reclaim. Tomorrow it'll be something else. Product names shift. Interfaces improve. Entire categories appear and disappear. What lasts is the inner operating system.

You begin to look at every new capability and ask different questions. - What becomes trivial now that used to consume me? - What becomes newly possible if I stop assuming the old constraints still apply? - What should I hand off? What should I protect? - Where's the smallest experiment that'd teach me something real this week? - If this capability doubles again, what kind of person do I want to already be?

Seen from the outside, none of this looks dramatic. Your calendar still has blocks. Your inbox still fills. Your week still has meetings, drafts, errands, obligations, and all the ordinary weather of adult life. But the texture changes.

- Monday begins with a one-page brief that Copilot drafted and you sharpened.
- Tuesday's meeting ends with clear action because Otter caught the decisions while the room was still warm.
- Wednesday gives you ninety uninterrupted minutes because the administrative floor held.
- Thursday includes one experiment that makes you uncomfortable in the useful way.
- Friday ends not with collapse, but with a short review: what to keep, what to kill, what to tune, what this week bought back.

You don't become a different person. You become the person you meant to be in a world where that shift finally matters.

The curve will keep rising. The tools will keep improving. Your job isn't to keep up with every shiny thing. Your job is to **aim the newfound capacity at a life that feels like yours**.

Part III – Health and Energy as the New Wealth

"It is health that is real wealth and not pieces of gold and silver."

– Mahatma Gandhi

7

AI for Health Habits and Recovery

Elena looked at the sleep summary on her phone and felt accused by it. The numbers were familiar: another short night, a higher resting heart rate than usual, and a recovery score that seemed to confirm what her body had already told her when the alarm went off. She was tired.

Nothing about the morning felt like an emergency. That was why it was so easy to dismiss. Elena could still get dressed, wake the children, answer a message from her manager, and make it through the first meeting. She'd become skilled at functioning below her best. Most people around her would've called that discipline.

By late afternoon, discipline looked different. She reread the same paragraph three times, reached for food without being hungry, and shortened a reply to a colleague because patience required energy she no longer had. Then she promised herself she'd go to bed earlier, exercise tomorrow, plan better meals on Sunday, and somehow become more consistent once the week became less demanding.

The week never became less demanding.

Elena didn't lack information about health. She knew sleep mattered. She knew movement mattered. She knew stress, food, and recovery shaped how she felt. Her problem was the distance between knowing those things and building a life in which they could happen repeatedly. Advice arrived as separate instructions. Real life arrived all at once.

That distance is where AI can become useful. Not as a doctor in your pocket, not as a machine that turns your body into a score, and not as a shortcut around qualified care. Its most responsible role is more modest: helping you organize signals, prepare better questions, translate professional guidance into a workable routine, and reduce the friction between a healthy intention and an ordinary Tuesday.

Your health is the infrastructure beneath everything else in this book. Attention depends on it. Emotional steadiness depends on it. The ability to learn, create, earn, lead, and care for other people depends on a body that can carry the life you're asking it to hold.

The goal isn't perfect optimization. It's a body you can listen to, habits you can live with, and a recovery system that helps you return.

Turn Signals Into Better Questions

Elena's first mistake had been treating each number as a verdict. A low score meant she'd failed. A better score meant she'd succeeded. That made the wearable emotionally loud and practically weak. Consumer health tools can be helpful for observing general patterns, but their estimates aren't the same as a diagnosis, and a single unusual reading rarely explains why a person feels the way they do.

So she changed the assignment. For two weeks, she collected only what she could gather without turning the day into a research project: approximate sleep and wake times, caffeine timing, movement, late meals, demanding workdays, and a one-sentence note about energy or symptoms. She included the trends from her wearable, but she stopped asking the device to tell her what they meant.

At the end of the two weeks, she used an approved AI assistant to prepare a summary she could review and, if useful, share with a clinician:

> ***"Organize the following two weeks of wellness notes and device trends into a one-page summary. Separate observations from possible explanations. Identify repeated patterns, missing information, and five questions I could discuss with a qualified health professional. Don't diagnose a condition, recommend medication changes, or treat a wearable estimate as a medical fact. If any note describes a potentially urgent symptom, tell me to seek appropriate medical care rather than analyzing it."***

The summary didn't announce a hidden answer. It showed that her hardest afternoons often followed late work, irregular meals, and short sleep in the same twenty-four-hour window. It also showed what the record couldn't establish. Stress might've contributed. Travel, illness, medication, or something the data never captured might've contributed too. The assistant could arrange the evidence. It couldn't determine the cause.

That distinction made the information more useful. Elena scheduled a routine appointment and brought a concise account instead of trying to remember a month of scattered details in the examination room. Her clinician could ask better follow-up questions because the timeline was visible. Elena could listen more carefully because she wasn't using half the visit to reconstruct what'd happened.

AI didn't replace the clinical conversation. It helped Elena arrive prepared for it.

This is one of the safest and strongest patterns for health-related AI. Use it to organize your own information, translate unfamiliar terms into questions, create a list for an appointment, or turn professional guidance into plain-language next steps. Keep diagnosis, treatment, medication, and urgent decisions with qualified people and regulated tools designed for those purposes.

The boundary matters most when fear is involved. A general AI system can sound certain while being wrong, overlook context it was never given, or produce an explanation that feels reassuring because it is fluent. New, severe, or worsening symptoms aren't a prompt-writing challenge. They deserve appropriate medical attention.

Privacy deserves the same seriousness. Health information can reveal more than the question you want answered. Elena removed names and details the assistant didn't need, used settings and products she understood, and avoided placing sensitive records into tools that weren't approved for them. She shared the minimum information required for the task.

A useful health system should make you better informed without making your most private information casually exposed.

Build Habits That Can Survive Real Life

The appointment gave Elena direction, but direction wasn't the same as a routine. Her clinician's recommendations were sensible: create a more consistent sleep window, build regular movement back into the week, eat in a way that supported steady energy, and leave room for recovery. Elena had heard versions of that advice before. The hard part was fitting it around school drop-offs, late meetings, travel, and the nights when someone in the house needed her.

In the past, she responded by building an ideal plan. The ideal plan contained early workouts, elaborate meal preparation, perfect bedtimes, and no evidence that Elena had ever met her own calendar. When Thursday broke the plan, the rest of the week usually followed.

This time she asked AI to design for the life she actually had:

"Using the schedule, preferences, equipment, budget, and clinician-approved guidance below, design a two-week minimum viable health routine. Include a regular sleep window, simple meals with shared ingredients, two short strength sessions, low-friction daily movement, and brief recovery periods. Give every habit a full version, a ten-minute version, and a version for a disrupted day. Don't add supplements, medical advice, or restrictions I haven't provided. Make the plan flexible enough to resume after a missed day without starting over."

The phrase minimum viable changed the plan. Monday's full workout was thirty minutes, but its disrupted-day version was one circuit with the dumbbells in her bedroom. Dinner didn't require seven new recipes. The assistant created four repeatable meals, reused ingredients, and grouped the shopping list by aisle. A late meeting didn't cancel the evening routine; it shortened it.

Her calendar became part of the health system. Two exercise blocks were scheduled where they had a reasonable chance of surviving. Walking calls replaced two meetings that didn't require a screen. A reminder at night didn't say Optimize sleep. It said, Put tomorrow's first decision on paper, plug in the phone outside the bedroom, and begin closing the day.

None of these actions was remarkable. Together, they removed negotiation from moments when Elena was least equipped to negotiate well.

A habit becomes powerful when the tired version of you can still find the next step.

AI is particularly good at this kind of translation. It can turn a food preference and a budget into a repeatable grocery list. It can adapt a professional exercise plan to the equipment available at home. It can find open space on a crowded calendar, prepare a packing list for travel, or build a reminder sequence around a routine you've already chosen.

What it shouldn't do is bury you under a new layer of maintenance. Elena didn't add six dashboards or record every bite. She chose a few signals connected to how she wanted to live: whether she protected the sleep window, moved most days, completed the two strength sessions, and felt more or less available by late afternoon. The system served the habit. The habit didn't serve the system.

By the third week, the largest change wasn't visible in any score. Elena no longer treated a missed workout or difficult night as evidence that the plan had failed. She opened the disrupted-day version, did what fit, and returned to the fuller routine when life allowed.

Consistency stopped meaning perfection. It began to mean returning without drama.

Treat Recovery as Part of the System

A month later, Elena entered the kind of week that used to erase every good habit. A deadline moved forward, one child came home sick, and two nights of broken sleep left her feeling heavy. Her wearable reflected the strain, but this time she didn't interpret the numbers as a challenge to push harder.

She treated recovery as information.

The old version of Elena would've tried to preserve the entire plan, then felt guilty when her body refused to cooperate. The new system asked a better question: What does taking care of the week require now? She reduced the workouts, protected the sleep opportunity she could control, simplified meals, moved nonessential work, and followed the guidance she'd already received about when to rest or contact a professional.

At the end of the week, she used a short review:

> ***"Review the wellness plan and notes below as a recovery coach, not a medical provider. Identify what supported energy, what added strain, and which commitments should be restored, reduced, or paused next week. Recommend one small experiment and one question for a qualified professional if the pattern continues. Treat missing data as unknown, avoid diagnosis, and favor recovery and sustainability over maximizing output."***

The review helped her see that recovery wasn't the absence of progress. It was the part of progress that made another cycle possible. That idea changed the way she viewed sleep after travel, lighter training after illness, quiet after an emotionally difficult week, and the decision to ask for help before a problem became a crisis.

It also kept AI in its proper place. The assistant could compare notes, notice repetition, prepare a plan, and remind her of the boundaries she'd already chosen. It couldn't feel pain, examine her, understand every interaction among her health history and circumstances, or carry responsibility for a consequential decision.

The more serious the consequence, the more important it is that a qualified human remains in the loop.

This principle extends beyond illness. Training plans need room for adaptation. Nutrition plans need room for culture, pleasure, access, and family life. Sleep routines need compassion for new parents, caregivers, shift workers, and people whose circumstances don't fit a tidy template. A system that labels every imperfect day a failure may be efficient at measurement and terrible at helping a person stay well.

Elena's system improved because it became less controlling. It gave her a way to notice without obsessing, prepare without pretending certainty, and return without punishing herself. The numbers became signals again, not grades.

Several months after that Tuesday morning, Elena still had demanding weeks. She still stayed up too late sometimes. She still missed workouts. Health hadn't become effortless, and no honest system could promise that it would.

What had changed was the distance between noticing and responding. A signal could become a question. Professional guidance could become a routine. A disrupted day could become a smaller version of the plan instead of the end of it. Recovery could begin before exhaustion had to shout.

That's the real opportunity for AI in health: not to guarantee an outcome, but to help you build a more attentive relationship with the body carrying you toward it.

Let the machine organize the notes, reduce the planning friction, and help good intentions survive a tired week. Keep for yourself—and for the professionals you trust—the judgment, consent, context, and care that health requires.

Your body isn't a dashboard to perfect. It's the home from which every ambition in this book must be lived.

The better you learn to support it, the more of yourself you can bring to what comes next.

8

Protecting Attention and Cognitive Energy

At 2:43 on a Wednesday afternoon, Nora stared at the cursor blinking in the middle of a sentence she'd begun before lunch. The strategy brief was still open. So were her inbox, the team chat, two dashboards, a meeting transcript, and a browser tab she'd opened to answer a question she could no longer remember.

She'd been busy for nearly six hours. She'd approved a request, found an old file, answered seventeen messages, moved a meeting, joined another, and helped a colleague reconstruct a decision from the previous week. Very little of it had been pointless. Almost none of it had moved the brief.

Nora still had time. What she no longer had was the same mind she'd brought to the desk that morning.

That's how attention is usually lost. Not in one dramatic collapse, but in dozens of reasonable withdrawals. A message needs an answer. A meeting needs context. A choice that should've been settled returns because no one recorded it. Each demand is small enough to accept. Together, they spend the mental energy required to hold a difficult thought in place.

We often treat an open hour as if it's interchangeable with any other open hour. It isn't. Sixty minutes with a clear mind can produce work that three depleted hours can't. The scarce resource isn't merely time. It's cognitive energy: your ability to focus, retrieve, compare, decide, and remain with a problem long enough to understand it.

AI can protect that resource, but only if it is used for the right work. Asking a machine to fill every available minute with more output creates another source of noise. Asking it to reduce sorting, switching, searching, and reconstruction can give the mind a cleaner place to stand.

The goal isn't to think less. It's to spend less of your thinking on the friction surrounding the thought.

Nora didn't need a new personality or a perfect focus ritual. She needed a different operating system for how information reached her, where unfinished work lived, and which parts of the day were allowed to belong to something deeper than response.

Give the Day One Front Door

The next morning, Nora resisted the habit of opening communication tools in the order their icons appeared. Instead, she entered the day through one page. Her approved workspace drew from the calendar, project system, meeting notes, and the messages she'd marked for follow-up. It didn't show everything. That was the point.

Before she looked at the inbox, she asked for a briefing:

> ***"Using only the connected sources below, prepare a morning brief with three priorities, the decisions that require my judgment, deadlines or risks that changed since yesterday, and the people waiting on me. Link each item to its source. Separate urgent from merely recent, identify conflicts, and state what you couldn't verify. Don't add tasks simply because a message is unread."***

The brief showed eleven things that were new and only three that needed Nora. One project was waiting on a decision she'd already made; the meeting record contained the answer. A message labeled urgent was a status update with no request. The strategy brief remained the most important work of the morning, even though it was the least likely to send a notification.

That changed how she used the first ninety minutes. She closed the communication tools, placed the source documents beside the brief, and began while the shape of the argument was still visible to her.

The value of a front door is not the dashboard itself. It is the act of choosing what earns first access to your attention. Without that choice, recency masquerades as importance. The newest message feels more real than the quiet project whose consequence is larger but whose deadline doesn't flash.

Attention becomes easier to protect when priority is decided before interruption begins.

AI can help prepare that decision by clustering related requests, distinguishing information from action, comparing deadlines, and showing where the same issue appears in several places. But the criteria must come from you. A model doesn't know which promise matters most, which relationship needs care, or which risk the organization is willing to accept unless those priorities have been made explicit.

Nora wrote hers at the top of the page: protect the client decision due Friday, unblock the team, and preserve one block for the work no one else could do. The assistant could sort the day's inputs against those commitments. It couldn't choose the commitments for her.

She also stopped treating her inbox as a place where work should live. Messages became inputs to a system, not permanent storage for obligations. If a note required action, it moved to the task list with an owner and date. If it contained a decision, it moved to the project record. If it required nothing, it was allowed to end.

By the end of the week, fewer loops were floating in her head. She wasn't trying to remember who needed what while also trying to solve the problem in front of her. The day still contained interruptions, but they no longer arrived in one undifferentiated pile.

Use AI as External Working Memory

The second change began after a meeting in which three intelligent people remembered the same decision differently. Nora could feel the old pattern beginning: a follow-up call, a search through messages, and another hour spent reconstructing an agreement that'd already consumed an hour to make.

This time the transcript, decision log, and project page were connected. The assistant could identify what'd been decided, who owned the next step, and which question had been left unresolved. Nora reviewed the summary before it was posted because transcripts miss nuance and models can turn an ambiguous exchange into false certainty. Once verified, the record became shared memory instead of private recollection.

The same approach changed her research. Instead of saving documents into folders whose logic she would forget, she captured each source with a short note: what question it answered, what evidence it contained, and what remained uncertain. Later, when she returned to the strategy brief, she didn't ask the model to write from general knowledge. She asked it to work inside the evidence she'd chosen:

> ***"Search the connected project sources for material relevant to the question below. Return the strongest sup-***

porting evidence, the strongest conflicting evidence, prior decisions that constrain the answer, and gaps that still require research. Quote or summarize only what you can link to a source. If two sources disagree, show the disagreement rather than resolving it for me."

The response acted less like an oracle and more like a librarian who knew the collection. It resurfaced a customer interview Nora had forgotten, connected it to a metric in the quarterly review, and found a constraint buried in a decision log. It also marked one important claim as unsupported.

That last result saved more cognitive energy than a confident answer would've. Nora didn't have to discover the gap after building the argument around it. She could investigate while the cost of changing direction was still small.

A trustworthy cognitive system doesn't merely help you remember. It helps you see what you don't know.

External memory has another benefit: it reduces the residue work leaves behind. When a decision, promise, or idea has a dependable place to go, the mind doesn't have to keep rehearsing it for fear of losing it. Nora could end a meeting by confirming one sentence, one owner, and one date. She could close the laptop knowing where unfinished work would reappear.

The system needed boundaries. Not every conversation belonged in a transcript. Not every document could be placed in an AI workspace. Permissions had to follow the information, and people needed to know when a meeting was being recorded or analyzed. A memory system that ignores consent or exposes confidential material may reduce one kind of friction while creating a much more serious one.

Nora kept the design simple: approved sources, narrow access, visible links, human review, and a clear place to correct the record. The assistant could retrieve and connect. People remained responsible for what was true, what was private, and what became part of the organization's memory.

Protect the Work Only You Can Do

By Friday, Nora had recovered several hours. The temptation was immediate: fill them. Add another meeting. Clear more messages. Produce two more drafts because the first one had become easier.

That would've missed the point.

Recovered time has value only when it is given a purpose. Nora used part of it for a protected block on the strategy brief and part of it for a conversation that needed patience. She left small buffers between demanding meetings so that one difficult exchange didn't spill its emotional residue into the next.

Inside the focus block, she used AI differently. It no longer organized the perimeter of the work. It challenged the work itself:

> ***"Act as a rigorous reviewer of the argument below. Identify the three assumptions most likely to be wrong, the strongest reasonable objection, evidence that would change the recommendation, and any sentence that***

sounds more certain than the support allows. Don't rewrite the brief. Ask questions that force me to think more clearly."

The response irritated her in useful ways. One assumption relied on a small set of customers. Another confused what was easy to measure with what mattered. A sentence near the conclusion sounded decisive because Nora was tired of uncertainty, not because the evidence justified confidence.

She revised the argument herself. That mattered. If she'd asked for a polished replacement, she might've received cleaner language without stronger reasoning. By asking the system to expose weakness, she kept ownership of the judgment while gaining another angle on it.

The best cognitive partnership doesn't remove productive difficulty. It removes avoidable difficulty so you can stay with the part that deserves you.

There are limits no workflow can erase. AI can't replace sleep, movement, quiet, or a nervous system given time to settle. It can't make six hours of intense meetings cognitively free. It can't protect attention while every notification remains invited and every recovered minute is immediately sold to another obligation.

So Nora made a few decisions the software couldn't make for her. The phone stayed outside the focus block. Some messages waited. The calendar held a real stopping point. Work that required a fresh mind moved earlier when possible, and low-stakes administration moved to the part of the day when depth was less available.

At 5:30, she read the strategy brief once more. It wasn't perfect, but it was coherent, supported, and clearly hers. More important, she could still feel the difference between finishing and merely stopping. The work had reached a natural edge. It wasn't following her home as a cloud of unnamed obligations.

That evening, Nora listened when someone spoke to her. She didn't keep one corner of her mind inside the browser. The benefit would never appear on a productivity dashboard, but it was part of the return.

Protecting attention isn't only about producing better work. It's about preserving enough of your mind to remain present for the rest of your life.

Let AI prepare the brief, retrieve the source, capture the decision, and challenge the weak assumption. Let it carry the mental lifting that doesn't require a human soul. Then spend the energy it returns with intention.

Your attention is one of the clearest expressions of what you value. Where you place it, repeatedly, becomes the shape of your days.

9

Emotional Intelligence in a Digital Age

Maya watched three words appear in the team chat: "Fine. Ship it."

The message came from Luis, the product lead who'd argued that the release wasn't ready. Ten minutes earlier, a senior executive had ended the debate by choosing the date. Now the channel had gone quiet. The words could've meant agreement, resignation, anger, or simple exhaustion. The screen delivered the sentence and removed nearly everything Maya would've used to understand it in a room.

Her first impulse was to answer quickly. She began typing, We all need to align once a decision is made. The sentence was defensible. It was also likely to turn a strained moment into a loyalty test.

Maya stopped.

That pause is where emotional intelligence begins. Not with perfect calm or an unusual gift for reading people, but with the ability to notice that your first interpretation is only an interpretation—and that stress is already trying to write the response.

Digital work makes this harder. Messages travel without tone. Silence becomes easy to misread. A clipped sentence can be reread until it feels sharper than the sender intended. AI adds another complication: it can generate language that sounds patient, warm, and emotionally fluent even when the person sending it hasn't done the human work behind those words.

That doesn't make AI useless in relationships. It changes the role it should play.

AI can help you slow down, examine your assumptions, prepare for a difficult conversation, and remove avoidable friction around care. It can't feel on your behalf, take responsibility for your impact, or counterfeit the attention another person deserves.

Maya deleted the reply. She sent Luis a private note instead: I may be reading too much into the channel, but I want to understand how that decision landed. Do you have ten minutes tomorrow?

The conversation that followed changed more than the message ever could have.

Read Before You Interpret

Maya began the next morning by reviewing what she actually knew. Luis had raised two specific quality risks. The executive had chosen to keep the date. The meeting had ended without deciding who would monitor those risks after launch. Luis's final message had been brief. Everything else—whether he felt ignored, embarrassed, angry, or ready to leave—was a story Maya's mind had added.

She used the approved meeting assistant to revisit the transcript and prepare questions, but she gave it a boundary:

> ***"Review the meeting notes and decision record below. Separate observable facts from my interpretations. Identify unresolved questions, moments where a concern may not have received a clear response, and information I should verify directly. Help me prepare five open questions for a private conversation. Don't infer a person's emotion, intention, mental state, or character from tone or word choice."***

The assistant showed that Luis's first risk had been discussed and accepted, but the second had disappeared when the conversation shifted to schedule. It also showed that no owner had been assigned to the monitoring plan. That didn't prove how Luis felt. It gave Maya something concrete to ask about.

In the conversation, she began with the gap rather than the guess. "I noticed your second concern didn't get a clear response before we moved to the date. What did we miss?"

Luis didn't say he was angry. He said he was worried the team would be blamed for a failure it had predicted and hadn't been given time to prevent. The distinction mattered. If Maya had entered the call trying to manage his attitude, she would've defended the organization against a feeling he hadn't named. By asking about the unresolved work, she reached the source of the tension.

She reflected before solving: You want the risk and the ownership to be visible before we launch.

Luis nodded. The room changed.

People often become easier to work with after they feel accurately understood, not because agreement has been manufactured, but because they no longer have to fight to prove that the problem exists.

The two of them created a short monitoring plan and took it back to the team. The release date stayed. The concern didn't vanish, but it had a name, an owner, and a response. Emotional intelligence hadn't removed the disagreement. It had made the disagreement usable.

AI can support this kind of listening by retrieving commitments, showing who spoke and who didn't, identifying questions that remained unanswered, or summarizing a long exchange before a follow-up. It should be much more cautious about claiming to know how someone feels. Inferred emotion is uncertain, culturally sensitive, and easy to weaponize when it is treated as a fact.

Maya adopted a simple rule: use the system to find the missing context, then ask the person for the meaning. A model could flag a place that deserved attention. The human being remained the authority on their own experience.

Use AI as a Mirror, Not a Mask

A week later, Maya received a message that made the lesson personal. A colleague wrote that her team had changed a requirement without warning and left another group to absorb the rework. Maya felt the accusation before she processed the facts. Her shoulders tightened. She drafted a response explaining why the other team had misunderstood the process.

Before sending it, she read the draft aloud. It sounded controlled. It also sounded like a closed door.

She asked AI to act as a mirror:

> ***"Review the draft below for defensiveness, sarcasm, absolutes, hidden accusations, and language that may make the other person feel dismissed. Preserve the facts, the boundary, and the decision I need to communicate. Suggest a more direct and respectful version that acknowledges the specific impact without pretending I know how the recipient feels. List any claim I should verify before sending."***

The assistant highlighted phrases such as "as we already explained" and "you should've known." Neither phrase advanced the decision. Both protected Maya's pride. It also noted that one timeline in her draft conflicted with the project record.

Maya corrected the fact and rewrote the opening: "We changed the requirement on Tuesday and didn't make the downstream impact clear. Your team had to redo work because of that. I want to confirm the new boundary and agree on how changes will be communicated from here."

The message wasn't softer in the sense of becoming vague. It was more accurate. Maya still explained the business reason for the change and held the new requirement. She also owned the part that belonged to her team.

Good emotional communication isn't the absence of a hard message. It's a hard message delivered without unnecessary injury.

This is where AI can be genuinely helpful. A person under pressure often can't hear the edge in language they wrote to defend themselves. A tone review can reveal blame, overstatement, or ambiguity before those words become another person's problem. A role-play can surface reasonable objections. A summary can help you separate the actual issue from the history you've attached to it.

The risk is using polished language as a substitute for honest intent. A generated apology isn't meaningful if you haven't accepted responsibility. A warm performance review isn't humane if the employee is still surprised by the decision. A message that says I understand may create distance if you never asked what the other person was experiencing.

Maya kept one standard for AI-assisted communication: she had to be able to say every sentence in the room. If a phrase sounded compassionate but didn't reflect what she believed or intended to do, it didn't belong under her name.

Let AI help remove the static. Don't let it impersonate the signal.

Over time, her difficult messages became shorter. She named the observation, explained the impact, asked what she might be missing, and made the next decision clear. When repair was needed, she did it directly. When a boundary was needed, she didn't hide it beneath artificial warmth.

Her colleagues didn't agree with her more often. They trusted the disagreement more.

Automate the Logistics of Care

The team still had a systems problem. Too much emotional friction was being created by ordinary operational failures: decisions vanished, people learned about changes late, one time zone carried every inconvenient meeting, and difficult customer cases remained inside automated loops long after a person should've stepped in.

Maya realized that empathy isn't expressed only in what a leader says. It is also expressed in how work is designed.

A clear decision record tells people they don't have to rely on influence or memory to know what was agreed. An owner and a date prevent the quiet resentment of work that belongs to everyone and therefore to no one. A buffer after a difficult conversation gives the nervous system time to settle. An early human handoff tells a frustrated customer that efficiency won't be valued above dignity.

Maya asked the team to redesign one high-friction workflow:

> ***"Design a responsible workflow for the situation below. Let AI handle routine intake, summarization, routing, and retrieval. Define the points where a person must take over, including repeated confusion, distress, conflict, sensitive information, or a consequential decision. Include consent and privacy safeguards, a way to correct the record, clear ownership, expected response times, and a brief explanation users can understand. Don't use inferred emotion as an automatic verdict."***

The resulting process was faster, but speed wasn't its most important feature. A customer no longer had to repeat the same story after a handoff because the verified context traveled with the case. An employee could challenge an inaccurate summary. A manager saw when a conflict had moved beyond routine resolution. Everyone knew which decisions automation could prepare and which decisions required a human name.

The team also became more careful with emotional data. A system that claims to detect frustration, disengagement, or stress can easily cross from support into surveillance. People may change how they speak when they know a machine is evaluating tone. A manager may treat a weak signal as a judgment about commitment. A model may misunderstand language shaped by culture, disability, personality, or context.

So Maya set a high bar: collect the least data necessary, tell people what the system is doing, make participation and correction meaningful, and never use an inferred emotional state as the sole basis for a decision affecting someone's work, opportunity, or dignity.

The more intimate the signal, the stronger the obligation to handle it with consent, humility, and restraint.

Several weeks after the Fine. Ship it. message, Maya's team held a review of the release. The product had launched with one manageable issue and one risk that never materialized. Luis explained what the monitoring plan had allowed the team to see. Another colleague admitted she'd stayed quiet in the original meeting because the debate had felt settled before she was ready.

Maya didn't ask AI to score the emotional health of the room. She asked the people what the process had made easier or harder. The assistant captured the decisions, grouped repeated themes, and prepared the follow-up. The meaning came from the conversation.

That is the future worth designing. Machines can carry more of the recall, routing, drafting, and administrative weight that often leaves people too depleted to treat one another well. They can help us notice a missing voice, reconsider a reactive sentence, and make the mechanics of follow-through more dependable.

But they can't decide who we become when the message lands badly, the deadline slips, the customer is afraid, or the person across from us needs more than an efficient answer.

As artificial intelligence becomes more capable, emotional intelligence becomes more consequential. Speed raises the cost of a careless response. Automation increases the value of genuine presence. Fluent machines make honest human attention easier to recognize.

Use AI to prepare for the conversation, examine the edge in your words, preserve the context, and make care easier to carry through the system. Then show up yourself.

Read before you interpret. Pause before you react. Ask before you assume. Repair when your impact doesn't match your intention.

The tools will keep getting better at sounding human. Our responsibility is to keep getting better at **being** human.

Part IV – AI in Business

"The human spirit must prevail over technology."

– Albert Einstein

10

The AI Advantage in Business and Career

Priya stared at a blinking cursor and a Slack thread that'd grown while she slept. A competitor had released the feature her team kept promising for "next quarter." Two customers had already asked whether something similar was on the roadmap, and a sales leader had forwarded the announcement with a single question: How fast can we respond?

Priya knew the old answer. She'd form a working group, request a market scan, ask Operations for updated customer data, and schedule a review for Friday. By the time the team had assembled the facts, the conversation would've moved on. The process was familiar and defensible. It was also too slow for the moment in front of her.

For a few seconds, she felt the pressure to match the competitor immediately. Then she recognized a more important distinction. Her team didn't need to copy a feature. It needed to understand what customers were actually asking for, whether the new release solved that problem, and what could be learned before the company committed months of work.

Priya opened the approved AI workspace connected to the company's support tickets, CRM records, product notes, and live reporting. The connections mattered. She wasn't pasting confidential customer information into a public chat or asking a model to invent a market answer from memory. She was giving the system access to sources the team already trusted, with permissions the company could explain.

Then she asked a narrower question than the one in the sales email:

> ***"Compare the competitor's announced feature with the recurring needs of our three largest customer segments. Cluster the relevant themes from support tickets and recent sales notes, link each conclusion to its source, and identify three small tests we could run this week. Rank the tests by likely customer value, effort, and risk. Flag missing or conflicting evidence instead of filling the gaps."***

The response didn't make the decision. It changed the shape of the decision. The loudest customer comments weren't necessarily the most common ones. A frustration that'd looked like a missing feature was often a confusing step in the existing workflow. One customer segment cared about the competitor's announcement; another cared far more about reliability and reporting. The team rejected one recommendation because it relied on too little evidence, refined a second, and chose a third as a small test they could place in front of customers that week.

By 10:15, Priya hadn't solved the market. She'd done something more useful: she'd shortened the distance between a signal and a responsible response. What once would've become a Friday presentation had become a Monday experiment.

That's the AI advantage in business and career. It isn't the ability to produce more material at greater speed. It's the ability to **see sooner, learn faster, and direct human effort toward the part of the problem that matters**.

The Three Edges of AI Advantage

The first edge is speed, but speed is easy to misunderstand. The goal isn't to make every person move faster or to turn each day into a race against a machine. The useful kind of speed removes waiting that never improved the work in the first place.

Priya's team had spent years accepting delay as a natural part of decision-making. Customer feedback lived in one system, sales notes in another, and operating data in a third. An analyst had to gather the pieces before anyone could begin thinking about them together. By the time the evidence became a slide, much of the discussion was spent arguing over whose version was current.

AI reduced that delay by preparing the first synthesis. It could cluster repeated language, compare segments, summarize a long thread, and show where two sources disagreed. That allowed the analyst to begin with interpretation rather than collection. The saved time didn't eliminate the analyst's role. It moved the role closer to judgment.

The second edge is cost, especially the cost hidden inside ordinary friction. A delayed handoff, an invoice checked three times, a new employee searching for the current procedure, or a customer repeating the same information to three people may not appear as a separate line in the budget. The cost is still real. It shows up as rework, lost patience, slower service, and capable people spending their attention on tasks that don't need their full ability.

When AI handles a first draft, prepares a reconciliation, routes a request, or turns meeting notes into clear owners and dates, the benefit isn't merely fewer minutes. The benefit appears when those minutes are deliberately returned to something more valuable: a better product decision, a stronger customer conversation, a careful review, or work that'd remained unfinished because administration always arrived first.

The third edge is capability. This is where the change becomes personal. A person can now approach work that once required a larger team or a longer chain of specialists. A sales manager can examine patterns across calls before deciding where coaching is needed. A designer can explore several visual directions before investing hours in one. An operations leader can model a workflow and test likely bottlenecks before changing the live process. A project manager can turn scattered updates into a decision brief and spend the meeting resolving trade-offs instead of reading status aloud.

Capability doesn't mean expertise has become unnecessary. It means **expertise can begin at a higher level**. The tool can help a newer employee understand the vocabulary of a problem; an experienced employee can use the same tool to compare edge cases, challenge assumptions, and move more quickly toward the judgment only experience can provide.

These three edges reinforce one another. Faster synthesis lowers the cost of testing an idea. Cheaper tests create more opportunities to learn. More learning expands what a person or team can do next. Over time, the advantage isn't one dramatic breakthrough. It's a compounding difference between an organization that learns from each cycle and one that repeatedly rebuilds the same understanding.

Priya made that difference visible by changing one question in the weekly review. Instead of asking, "What did we finish?" she began asking, "What did we learn soon enough to change?" The room became less interested in the volume of activity and more interested in whether the activity had improved the next decision.

Multiply the Work, Not the Noise

Once the first experiment was underway, Priya looked at how the team itself was working. She didn't begin by buying more tools. She asked each person where effort was being spent without improving the outcome.

The support lead pointed to the hours required to read hundreds of comments before a pattern could be named. The analyst described the time lost cleaning exports that changed format every month. The product manager admitted that she rewrote the same update for executives, engineers, and customer teams because each audience needed a different level of detail. The sales leader said his team had more account data than ever and less confidence about which signals deserved attention.

None of these people wanted a machine to take over the work they were proud of. They wanted relief from the layers around that work.

Together, they ran a role-level friction audit:

> ***"Review the responsibilities below as if you were helping redesign this role. Separate the work into four groups: repetitive preparation, pattern-finding, human judgment, and relationship or accountability work. Suggest where an approved AI tool could assist with the first two groups, what evidence a person should verify, and what should remain fully human. Don't recommend removing the role. Focus on helping the person contribute at a higher level."***

The exercise produced a useful boundary. AI could prepare, organize, compare, and propose. People would decide, approve, explain, and carry the consequences. That boundary was simple enough to remember when a new use case appeared.

The support lead began receiving a weekly cluster of recurring issues, with links back to representative tickets. She no longer read every ticket merely to discover that the same confusion had appeared fifty times. She used the recovered time to listen closely to the conversations where emotion, complexity, or risk made categorization unreliable.

The analyst automated part of the data preparation, but not the controls around it. He reviewed record counts, missing fields, unusual changes, and the logic behind every calculated measure. Instead of being valued as the person who could endure the spreadsheet, he became the person who could explain what the numbers meant and when they shouldn't be trusted.

The product manager created one source brief and used AI to adapt it for different audiences. She still reviewed every version because compression can remove the caveat that matters most. The point wasn't to generate three times as much communication. It was to stop rewriting the same facts from a blank page and spend more attention on whether each audience understood the decision.

The sales team treated scoring as a reason to investigate, not a verdict. A high-risk account prompted a review of call notes, product usage, unresolved issues, and the customer's own words. The score directed attention. It didn't replace the relationship.

This is the career opportunity beneath the fear of automation. Roles are rarely bundles of identical tasks. They contain repetition, interpretation, coordination, judgment, and trust in different proportions. AI may absorb or reshape some tasks while increasing the value of the rest. The person who learns to manage that boundary becomes more useful, not because they can produce endless drafts, but because they can combine machine throughput with context, taste, and responsibility.

The strongest career question is therefore not, “Will AI replace my job?” It is, **“Which part of my work is becoming cheaper, and which part becomes more valuable because of that?”**

If summaries become cheap, the ability to identify the decision becomes more valuable. If first drafts become cheap, taste and revision matter more. If data access becomes conversational, asking a precise question and recognizing a misleading answer matter more. If routine service becomes automated, the human who can handle ambiguity, frustration, and repair becomes more important at the exact moment the interaction becomes difficult.

This shift requires learning, but not constant tool-chasing. Priya asked the team to choose one approved assistant, one reliable place for shared knowledge, and one automation platform where needed. Each tool had to solve a named problem. If it created another inbox, another dashboard, or another place to check, it had to earn that burden.

The team also kept a short Friday ritual: keep, change, or remove. Which workflow saved time without lowering quality? Which prompt produced confident but shallow work? Which automation failed quietly? What human task received the time that’d been recovered? The review kept the system connected to outcomes instead of novelty.

The result wasn't a workplace filled with people talking to machines all day. It was a workplace with fewer avoidable handoffs, clearer decisions, and more room for people to do the work that made their roles worth having.

Scale with Trust

The first week of the pilot moved quickly enough that Priya could feel a new risk forming. Success creates its own kind of carelessness. Once people see a useful result, they naturally want to connect more data, automate more steps, and move from suggestion to action. The question changes from "Can this help?" to "How far can we let it go?"

Priya's legal partner brought the team back to earth with a question written across the whiteboard: **Who could be affected if this system is wrong?**

That question slowed the conversation in the right way. A draft marketing headline carried little consequence and could be reviewed before publication. An account-risk score could influence how a customer was treated. A hiring recommendation, a credit decision, or a health-related flag could alter a person's opportunity or care. The more a system touched money, rights, safety, employment, or dignity, the more explanation, testing, and human authority it required.

The team designed the next pilot around those stakes:

"Create a limited pilot plan for the workflow below. Define the business problem, the people affected, the data required, the decisions the AI may support, and the decisions it may not make. Include a human approval point, a way to trace conclusions to evidence, tests for likely failure and uneven impact, a process for reporting errors, and clear conditions for pausing or ending the pilot."

The plan didn't make risk disappear. It made responsibility visible. Permissions were narrow. Logs showed what the system had done. Customer-facing messages required approval. No important record could be changed and no money could be committed without a person reviewing the action. The team tested ordinary cases and uncomfortable ones, including missing data, contradictory instructions, unusual customer language, and requests designed to push the system beyond its role.

They also told employees what the pilot was for. Priya didn't promise that every role would remain unchanged; no honest leader could make that promise. She did explain what the team was trying to remove, what skills would become more important, what information the system could see, and how people could challenge a poor result. Skeptics were invited to find the cracks because the people closest to a process often know where an elegant design will meet a difficult reality.

Trust grew less from reassurance than from participation. When the support lead showed that the model repeatedly misunderstood a phrase used by one customer group, the team didn't dismiss the concern as resistance. It corrected the workflow and added a test. When the analyst found a data gap, the launch slowed until the gap could be understood. Those moments taught the organization that governance wasn't a ceremony performed after the exciting work. It was part of making the work dependable.

There was another form of trust to protect: the customer's sense of being treated as a person. Efficiency can become emotionally expensive when every interaction feels processed. Priya's team wrote a rule for automation: routine should feel easy, but difficulty should make a human easier to reach. If a customer repeated the same problem, expressed distress, disputed a consequential decision, or entered a situation the system didn't understand, the handoff happened early. The AI could carry context into the conversation so the customer didn't have to begin again. It couldn't counterfeit care.

Several weeks after the competitor announcement, Priya reviewed what had actually changed. The team hadn't copied the feature. Its small experiment showed that customers valued a clearer workflow more than a larger set of options. Support had gained a faster way to surface themes. Product decisions were tied more closely to evidence. Employees had learned where AI strengthened their work and where it needed firmer boundaries.

The most important change was harder to place on a chart. The team felt calmer. Speed no longer meant frantic motion because less energy was being spent gathering, rewriting, and rediscovering. People could see the signal, understand the choice, and know who was responsible for the next step.

Priya wrote four promises beneath the question on the whiteboard:

We shorten the distance between a signal and a responsible response.

We direct the time we recover toward work that matters.

We help people grow into higher-value work instead of leaving them behind.

We keep a human name attached to every consequential decision.

The promises weren't a technology strategy. They were a way of working. The market would keep moving, tools would keep changing, and no team would remain ahead forever. But Priya's organization had learned how to respond without surrendering judgment, trust, or the people whose knowledge made the response intelligent in the first place.

That's the advantage worth building. **Automation forms the floor. Humanity sets the ceiling.**

11

Wealth Creation in the AI Economy

At 9:47 on a Sunday night, Hana sat at her kitchen table with three browser windows open and no real sense of what any of them meant together. One showed her retirement account. Another held a spreadsheet of household expenses she hadn't updated in two months. The third displayed a market chart that'd climbed sharply all week, accompanied by a headline suggesting that anyone who wasn't buying now was already behind.

She had a stable job, a small emergency fund, and an idea for a side business built around a problem she understood well. On paper, she was doing many of the things responsible adults were told to do. In practice, her financial life still felt scattered. She knew what had happened in each account, but she couldn't see what the pieces were building toward.

For months, Hana had heard that AI could help people create

wealth. Most examples sounded like shortcuts: a machine that'd find the next winning stock, launch a business over a weekend, or produce passive income while she slept. The claims were exciting precisely because they asked so little of the listener. They also made her uneasy. If building wealth were that automatic, why did every promise seem to end with someone selling access to the promise?

She closed the market chart and wrote a better question at the top of her spreadsheet: **What would it look like to use AI without handing it control of my financial life?**

The answer didn't arrive as a stock symbol or a business slogan. It came as three ordinary verbs: **create, keep, and grow**. Create more value in the world. Keep more of the money that value produces. Grow what remains through patient ownership and sound decisions. AI could support all three, but it couldn't remove risk, manufacture demand, or replace discipline. It could help Hana see more clearly, test more cheaply, and act more consistently. That was less glamorous than a shortcut, but far more useful.

Build the Money System Before You Chase the Money

Hana began with the part of wealth creation that rarely appears in a dramatic advertisement: knowing where the money already went. A personal-finance platform such as Monarch can organize connected accounts, categorize transactions, and answer questions about the picture those accounts create. Copilot Money can help clean up transaction categories and make spending patterns easier to review. A private spreadsheet can work too, especially when account names and other identifying details are removed before the numbers are shared with a general AI assistant.

The tool mattered less than the habit. Hana gathered twelve months of income and expenses, corrected the categories the software had guessed incorrectly, and separated the household into three layers. The first contained commitments she had to carry: housing, insurance, transportation, utilities, and minimum debt payments. The second contained choices that could change: subscriptions, dining, shopping, travel, and the many small conveniences that'd become invisible through repetition. The third contained the future: savings, retirement contributions, education goals, business investment, and the irregular costs that always seemed to arrive as surprises even though most of them arrived every year.

Once the categories were honest, she asked for help interpreting them:

"Using the monthly figures below, organize my cash flow into fixed commitments, flexible spending, irregular expenses, debt payments, and future-building contributions. Identify the three patterns that deserve my attention. Then model three conservative scenarios for the next twelve months: maintaining the current plan, redirecting a modest amount toward the emergency fund, and redirecting that same amount toward debt. Show every

assumption, don't choose a scenario for me, and list the questions I should review with a qualified financial professional."

The first response wasn't profound, but it made several quiet leaks visible. Two annual subscriptions had renewed without being used. A quarterly insurance bill was missing from the monthly view. Restaurant spending wasn't catastrophic, but it rose reliably during the weeks when work crowded out meal planning. The model didn't tell Hana that she'd failed. It showed her where the system was asking willpower to solve a design problem.

She canceled what no longer served the family, created sinking funds for the irregular costs, and moved an automatic transfer to the day after payday. None of those changes made her wealthy by Tuesday. Together, however, they increased the amount of money that could begin working toward something instead of disappearing into the space between intentions.

This is the first personal use of AI for wealth: **turning financial fog into a decision you can name**. A model can compare a higher retirement contribution with faster debt repayment. It can show how a raise, childcare change, vehicle purchase, or interest-rate shift might affect the next year. It can help a couple prepare questions for an advisor without spending the first half of the appointment trying to remember the facts. It can also challenge an optimistic plan by adding taxes, maintenance, inflation, fees, and the emergencies people tend to leave out when they want an idea to work.

For a household, the useful dashboard is often simple: cash on hand, monthly surplus or shortfall, high-interest debt, progress toward near-term reserves, long-term contributions, and the few goals that matter most. For a business, the same principle expands to revenue, gross margin, operating expenses, unpaid invoices, taxes set aside, customer concentration, and the cash needed to survive a slow month. QuickBooks and similar accounting platforms increasingly use AI to categorize transactions, surface cash-flow patterns, and support invoicing and follow-up. Those capabilities can reduce blind spots, but the books still need review, and tax decisions still belong with a qualified professional.

Hana adopted a rule before connecting another account: **convenience doesn't outrank privacy**. She reviewed what each tool could access, what it stored, whether its AI features were optional, and how the data could be deleted. She never pasted account numbers, tax identifiers, unredacted statements, passwords, or authentication codes into a general chatbot. When the stakes increased, she used AI to prepare—not to authorize. The machine could organize the question; it couldn't move the money.

By the end of the evening, Hana hadn't discovered a secret source of income. She'd built something more durable: a financial baseline she trusted. For the first time, the number in her investment account was connected to the rest of her life rather than floating above it like a separate score.

Track Markets Without Chasing the Noise

The next morning, the same stock-market headline was still waiting for Hana. This time she noticed what it didn't contain. There was no discussion of the company's cash flow, debt, competitive position, valuation, or the price an investor was being asked to pay for future expectations. The headline had urgency, but very little context.

AI can be genuinely useful in investment research because markets produce more information than any individual can read: annual reports, quarterly filings, earnings-call transcripts, investor presentations, economic releases, price movements, and breaking news. The opportunity isn't to ask a chatbot, 'What stock will go up?' The opportunity is to build a research process that notices meaningful changes and carries the reader back to the evidence.

Hana created a small watchlist rather than an endless feed. TradingView watchlist alerts could notify her when a price or other selected condition changed. Fiscal.ai could help her explore structured company data and ask research questions conversationally. Her brokerage provided screeners and research reports. Most important, the SEC's EDGAR database gave her the original filings for U.S. public companies. Each tool served a different role: **alerts found the signal, AI organized the reading, and primary sources settled the facts**.

She asked her assistant to prepare a weekly brief, but she gave it boundaries:

"Create a source-linked weekly brief for the companies and funds on this watchlist. For each one, summarize new SEC filings, earnings releases, material changes in revenue, margins, free cash flow, debt, share count, and management guidance, along with major news that could affect the investment thesis. Separate reported facts from interpretation, identify conflicting information, include publication dates and direct source links, and say when current data is unavailable. Don't recommend a trade or predict a price."

That prompt changed the work. Instead of receiving a confident paragraph built from whatever the model happened to recall, Hana received a reading list. One company on her watchlist had posted an earnings 'beat,' and the price jumped before the market opened. The brief also pointed her to the quarterly filing, where she saw that margins had narrowed, receivables had risen, and management had changed the language around demand. None of those facts automatically made the company a poor investment. They did make the celebratory headline incomplete.

Hana wrote four lines beneath every investment she researched. **The thesis** explained why the asset might create value over time. **The evidence** listed the facts that supported that belief. **The risks** named what could make the thesis wrong. **The price** asked how much optimism was already reflected in what she'd have to pay. AI was good at helping her compare those lines across filings and time. It was also capable of inventing a figure, overlooking a footnote, using stale data, or confusing a similarly named company. Every important number returned to the source before it entered her decision.

The same process can support a long-term investor who owns diversified funds rather than individual companies. AI can compare expense ratios, holdings overlap, asset allocation, historical drawdowns, and the tax location of different account types, provided the underlying data is current and verified. It can prepare questions about rebalancing or retirement-plan options. It can't know how a real person will respond when the market falls, whether a family can tolerate a loss at the wrong time, or how a change fits into that person's complete tax and financial picture.

Market tracking becomes dangerous when monitoring turns into stimulation. A dozen price alerts can make a long-term plan feel like an hourly emergency. A model that produces a new opinion every time it's prompted can give anxiety the appearance of research. Hana therefore chose conditions that could change her thesis, not merely her mood: a new filing, a material change in guidance, an unusual balance-sheet movement, a valuation entering a range she'd already decided to study, or a scheduled portfolio review. She didn't need AI to help her react faster. She needed it to help her react less often and for better reasons.

She used the same discipline beyond stocks. For a possible rental property, she asked the model to test the deal under conservative assumptions for vacancy, repairs, insurance, taxes, management, financing, and a major unexpected expense. For a small-business investment, she compared the cost of equipment or software with the realistic cash it could return and the time required to operate it. For a career decision, she modeled the value of a certification, a job change, or a reduced commute without pretending every benefit could be known in advance. In each case, AI widened the range of scenarios she could examine. It didn't remove uncertainty from the world.

Investing involves risk, including the possible loss of principal. AI-generated financial information can be inaccurate, incomplete, or outdated. The examples in this chapter are educational, not individualized investment, tax, legal, or accounting advice. Important decisions should be reviewed against current primary sources and, when appropriate, with a qualified fiduciary, tax professional, attorney, or accountant who understands your circumstances.

The most valuable result of Hana's research system wasn't a winning prediction. It was a calmer relationship with information. She could follow the market without allowing every movement in the market to enter her nervous system. **AI had become a research desk, not a roulette wheel.**

Turn What You Know into Income and Ownership

A sound plan for keeping and growing money still depends on there being money to direct. That's where AI can affect wealth more immediately: not by printing income, but by helping a person create more value, reach the right customer, learn a valuable skill, negotiate from better information, or turn repeated knowledge into something that can be sold more than once.

Hana's idea had been following her home from work for years. Growing companies often kept their most important procedures in too many places: old documents, shared drives, chat threads, and the memories of a few experienced employees. New hires wasted days finding the right version. Managers answered the same questions repeatedly. Errors appeared not because people were careless, but because the system made the correct action hard to find. Hana knew how to organize that mess into clean, searchable playbooks. What she didn't know was whether the frustration was painful enough for a customer to pay to solve it.

Before building anything, she asked AI to help her examine the opportunity from the customer's side:

> ***"You have experience in operations, process design, and knowledge management. Identify five expensive or recurring problems that organizations may face in these areas. For each problem, describe who feels it most, what they may be doing today, what evidence would show the problem is real, what a small paid service could test, and why a buyer might say no. Don't invent market-size figures or customer quotes. End with ten interview questions that explore the problem without leading the person toward my solution."***

The response gave Hana possibilities, not proof. She took the questions into conversations with former colleagues and small-business owners. Several ideas that sounded clever on the screen went nowhere in the room. The procedure problem didn't. People described slow onboarding, duplicated work, outdated instructions, and key employees who could never take a full vacation because too much knowledge lived only in their heads. The language was more specific than anything AI had written because it came from people carrying the cost.

Hana built the smallest honest version of the service. An AI assistant helped structure the offer and draft a landing page, but she removed every outcome she couldn't yet support. A Tally form collected interest. A short Loom video showed a before-and-after example in her own voice. Make routed approved information into her workspace. The page didn't pretend she had a large company, proprietary technology, or guaranteed results. It named the problem, explained the pilot, and made the next step easy.

Her first paid engagement was deliberately manual. Hana interviewed the team, cleaned the source material, built the playbook, and watched where employees still became confused. AI transcribed interviews, clustered repeated questions, compared conflicting versions, and drafted consistent formats. Hana made the decisions about what was true, what was sensitive, what should be deleted, and how the final guidance should sound. By keeping the first delivery close, she learned the work before automating the work.

That distinction matters because speed alone doesn't create wealth. If AI helps a business produce more work nobody values, it's accelerated waste. If it allows a founder to promise more than the operation can deliver, it's accelerated distrust. Wealth begins when faster learning becomes a better offer, when a better offer becomes reliable cash flow, and when reliable delivery becomes an asset the owner doesn't have to rebuild from zero each time.

Hana began to see an income ladder. At the first level, AI improved **earned income**: it helped her learn, prepare, and perform work that the market valued. At the second, it supported **business cash flow**: a repeatable service solved a clear problem for a defined customer. At the third, it helped create **owned assets**: templates, training, licensed methods, a searchable knowledge base, or software that could continue producing value beyond a single hour of labor. Each level required more than a prompt. It required trust, evidence, distribution, and the willingness to improve after the first version met reality.

The same ladder can begin inside a career. A project manager can use AI to identify skill gaps across job postings, build a learning plan, and prepare evidence for a promotion conversation. A tradesperson can turn years of field knowledge into clearer estimates, maintenance guides, and a training program for new employees. A teacher can shape original lessons into a course or workbook. A designer can license a reusable system instead of selling only isolated deliverables. A specialist can begin with consulting, learn what clients repeatedly need, and then build a product around that pattern. The tool lowers the cost of exploring the idea. The person still has to become credible enough for someone to trust it.

For commerce, tools such as Shopify Sidekick can help a merchant examine store data, draft content, create reports, and build workflows inside the business the merchant already operates. For accounting and cash flow, QuickBooks AI features can support categorization, forecasting, invoicing, and payment follow-up. Zapier or Make can connect approved steps across a small operation. These tools are useful when they strengthen a real model. They can't rescue a product customers don't want, margins that never worked, or a founder who refuses to look at the numbers.

Hana gave herself a monthly wealth review with three pages. The first asked, **What value did I create?** She looked at skills gained, problems solved, customer outcomes, and revenue earned. The second asked, **What did I keep?** She reviewed household surplus, business profit, taxes reserved, debt, and unnecessary leakage. The third asked, **What now belongs to the future?** She tracked contributions, durable business systems, intellectual property, customer relationships earned honestly, and the choices those assets might make possible later.

Three months after that Sunday night, nothing about Hana's life looked like an overnight-success story. She hadn't discovered a perfect stock, launched a company that ran itself, or escaped the need to make careful decisions. She'd canceled expenses that no longer mattered, automated a contribution she could sustain, built a market brief that reduced impulsive checking, and completed two paid pilots for a service rooted in experience she already possessed. She understood her household more clearly, and her business idea had moved from imagination into evidence.

That's the more honest promise of AI and wealth. It can help you

ask better questions before money moves. It can make public information easier to examine, patterns easier to notice, experiments cheaper to run, and useful knowledge easier to package. It can help a small business protect margin and a family protect intention. It can expand the number of intelligent moves available to one person.

What it can't do is guarantee that an investment will rise, that a customer will buy, that a business will succeed, or that a plan will survive every change in life. Those outcomes still meet risk, timing, competition, behavior, luck, and the parts of the future no model can see.

Hana closed the three browser windows and left one sentence at the top of her dashboard for the month ahead: **Create value. Keep a share. Grow it with patience.**

The sentence wasn't a shortcut. It was a direction. And unlike the flashing headline that'd started the evening, it was quiet enough to follow for years.

12

The Democratization of Intelligence

The power flickered once in the small library, then steadied. Layla waited for the ceiling fan to find its rhythm before touching the record button on her phone. Across the table, her grandmother adjusted a floral scarf and began telling a story Layla had heard only in pieces: a wedding held behind blackout curtains, a market flooded after an unusual rain, and the sound bread made when it was ready to come out of the oven.

Layla had been trying to preserve these stories for years. She could record them, but turning hours of conversation into readable text took more time than she had. Translating the stories for relatives who no longer spoke the language fluently required help she couldn't always afford. Building an archive that people could search felt like the kind of project that belonged to a university, a foundation, or a team with technical staff.

That afternoon, the project became smaller enough to begin. A speech-to-text system created a rough transcript. A translation model produced a second version for Layla to review with two bilingual relatives. An AI assistant helped her organize names, places, and recurring themes into a simple index. None of the outputs was ready to publish. Some phrases were wrong, a family name had been mistaken for an ordinary word, and the translation made one joke sound strangely formal. But the blank archive was no longer blank.

Her grandmother watched Layla correct a sentence and asked whether the machine understood the story.

"Not really," Layla said. "It helps me carry it."

That distinction is the heart of democratized intelligence. AI doesn't give everyone expertise, erase the need for teachers, or make unequal access disappear. It lowers the threshold for beginning. It gives more people a way to ask, translate, compare, practice, draft, and build before they have permission from an institution or the resources to hire a team.

For most of history, useful knowledge has been shaped by proximity: proximity to a teacher, a library, a profession, a language, or the people who could open a door. The internet widened access to information, but access to information isn't the same as the ability to use it. A page can contain the answer and still leave a reader stranded. AI adds a conversational layer. A person can ask for another explanation, a slower example, a translation, a comparison, or a first plan connected to the situation in front of them.

The shift isn't from ignorance to instant mastery. It's from **permission to participation**.

From Permission to Participation

Layla saw that shift each Saturday as the library's oral-history table began attracting other people. A retired mechanic brought a box of photographs with dates written only on the backs. A teenager wanted to interview his grandfather but didn't know what to ask. A mother hoped to preserve recipes that existed as gestures and handfuls rather than measurements. None of them arrived calling the project "AI." They arrived because something mattered and they were afraid it might disappear.

The tools helped with the parts that had once made the project feel too large. Audio could become a draft transcript. Faded handwriting could be enlarged and compared. A list of names could be organized into questions for the next interview. Translation could give a family two versions to place side by side. A simple website builder could turn the growing collection into pages that relatives in other places could visit.

The work still required people. Someone had to know that two spellings referred to the same person. Someone had to recognize when a translated phrase was technically accurate and emotionally wrong. Someone had to ask whether a story belonged in a public archive or only within the family. The machine could widen the doorway. It couldn't decide what should pass through it.

The same pattern appears anywhere a person has been kept at the edge of a conversation by language, confidence, cost, or credentials. A student can ask for a math concept to be explained with a different metaphor and then work through a new problem without pretending to understand. A small-business owner can draft a message to an overseas supplier, compare translations, and enter the conversation with greater confidence. A patient can turn unfamiliar medical terms into a list of questions for a clinician without confusing preparation with diagnosis. A first-time founder can organize questions for an attorney or accountant and use limited professional time more wisely.

In each case, AI doesn't replace the expert. It helps the person become a more capable participant in the expert conversation.

That can be especially meaningful for people who've learned to hide confusion. Asking a teacher, manager, or professional to explain something again can feel exposing. A patient tutor that doesn't become impatient can give someone room to reveal the exact point where understanding broke. The private rehearsal can create the confidence needed to speak in the real room.

Layla used that idea when the teenager at the library admitted that he'd never interviewed anyone. Together, they asked an assistant to prepare a guide:

"Help a first-time interviewer record an older relative's life story. Create twelve open-ended questions that begin with ordinary sensory memories before moving toward difficult history. Add gentle follow-up questions, re-

minders to ask permission before sensitive topics, and a closing question that lets the storyteller decide what matters most. Don't invent details or assume the person wants every memory recorded."

The guide didn't conduct the interview for him. It gave him enough structure to listen without panicking about the next question. During the conversation, he ignored half the list because his grandfather mentioned a train station and began telling a story neither of them had expected. That wasn't a failure of the plan. It was the plan doing its job: carrying the beginner far enough that human curiosity could take over.

This is how participation grows. The person uses assistance to cross the first barrier, then develops the skill to notice what no template could predict.

When Access Becomes Economic Leverage

Wider participation does more than make information easier to reach. It allows local knowledge to travel farther and become more useful.

Layla's archive began as a family project, but the stories contained knowledge the neighborhood recognized: traditional cooking methods, descriptions of streets that no longer existed, accounts of migration, repairs, celebrations, and the ways people cared for one another during hard years. A local teacher asked whether a few recordings could support a history lesson. A cultural center wanted to include excerpts in an exhibit. A younger relative offered to build a map linking each story to a place.

Before accessible AI tools, each request would've created another expensive layer: transcription, translation, editing, tagging, web design, and accessibility. The tools didn't make those disciplines unimportant. They allowed Layla to prepare a credible first version, understand where specialist help was most needed, and direct a modest budget toward the work that required real expertise.

That's economic leverage at ground level. A person closest to a problem can test an idea before finding a large sponsor. A teacher can adapt an explanation to the language and examples a particular community understands. A tradesperson can turn years of hard-earned knowledge into clearer training for apprentices. A community organization can prepare multilingual information and spend more of its limited resources on direct service. A small producer can describe a product for customers in another market without pretending cultural fluency they don't possess.

The advantage comes from pairing broad tools with specific knowledge. AI knows patterns across a vast amount of material. The person knows this customer, this field, this classroom, this family, this neighborhood, and the exceptions that matter here. When those forms of knowledge meet, one person can contribute at a level that previously required more money, more time, or a larger organization.

Layla began to see a second possibility. The archive didn't need to become a large company to create value. It could support paid workshops on interviewing elders, offer a template for other community groups, or provide carefully licensed material to an exhibit. Any income could support better equipment, professional translation, and preservation. The work could remain mission-led while becoming sturdy enough to continue.

She asked the assistant to help her examine that possibility without turning heritage into a sales pitch:

> ***"I'm building a community oral-history archive with permission from participating families. Identify five ways the project could create sustainable value without exploiting the stories. For each option, explain who benefits, what consent or compensation may be needed, what work should remain with qualified people, and what could go wrong. Favor small tests and community ownership over rapid scale."***

The response gave her a set of possibilities, not a business decision. Some ideas felt appropriate; others didn't. The most useful result was the list of questions it forced into view: Who owns the recording? Can a family withdraw it later? Should revenue be shared? What context would be lost if a story were separated from the speaker? Could a translation change the meaning in ways the listener would never see?

Access becomes leverage only when the people whose knowledge creates the value remain visible in the arrangement.

This principle travels beyond community archives. AI can help a designer compare research, a farmer organize observations before speaking with local agricultural guidance, or a small manufacturer turn scattered operating knowledge into a training tool. But the tool is most powerful when it strengthens the person already connected to the problem. The future isn't only large organizations distributing intelligence outward. It's also people at the edges bringing lived knowledge inward and building from where they stand.

That doesn't guarantee equal outcomes. Devices, connectivity, language coverage, disability access, education, and confidence remain uneven. Some communities are richly represented in training data; others appear faintly or through someone else's description. Democratization isn't accomplished because a tool is available in an app store. It requires infrastructure, teaching, accessible design, fair pricing, and people willing to share what they learn.

Still, the threshold has moved. More people can make a first version, prepare a stronger question, and demonstrate what an idea could become before the traditional gatekeepers decide whether it deserves attention.

Access Comes with Responsibility

As the archive grew, Layla discovered that the easiest part of the project was creating more material. The harder part was deciding what deserved to become public.

One recording included a family disagreement that'd never been resolved. The transcript made it look clean and factual, but the speaker's pauses told a more complicated story. Another contained the name of a person who hadn't consented to be included. A translation softened a phrase that carried real anger in the original language. If Layla had trusted the system because the pages looked polished, she'd have published errors with the appearance of authority.

She created a review process that moved more slowly as the consequences increased. Every transcript returned to the storyteller or family representative. Every translation received a human review. Sensitive names were removed or approved. Public pages explained when AI had assisted with transcription or translation. Original audio remained connected to the text when permission allowed, so readers could hear the voice rather than treating the generated transcript as the complete truth.

Her process can be adapted to almost any use of widely accessible intelligence:

> ***"Review the project described below for participation, accuracy, and harm. Identify who's represented, who may be missing, what claims require a primary source or qualified expert, what personal data should be removed, and what decisions need direct consent. Then create a short human-review checklist that must be completed before the work is shared."***

The checklist matters because generated language can make unfinished thinking look finished. AI speaks smoothly even when the evidence is weak. Wider access to creation therefore requires wider habits of verification. A person who can produce a persuasive page in minutes also needs to learn how to trace a claim, compare sources, label uncertainty, and recognize when a request is outside the tool's competence.

Skill itself needs protection. If a student asks for every answer, the tool can remove the struggle through which understanding develops. If a writer accepts every first draft, fluency may grow while voice weakens. If a community uses translation without anyone maintaining the language, access can quietly become another form of dependence. The answer isn't to preserve difficulty for its own sake. It's to keep the difficulty that builds judgment while reducing the friction that keeps people from practicing at all.

Privacy creates another boundary. A story, health record, school assignment, business plan, or photograph may contain information about people who never agreed to place it inside an AI system. Permission should be understandable, specific, and reversible where possible. Collecting less is often safer than promising to protect more.

There's also the growing problem of synthetic deception. The same tools that help a person translate a grandmother's memory can imitate a voice, fabricate a quotation, or create an image that borrows the emotional authority of an event that never happened. Verification will become an everyday habit rather than a specialist skill. Consequential requests should be confirmed through another channel. Surprising media should be traced to an original source. Speed shouldn't decide what becomes belief.

Layla didn't let those risks end the project. She let them define the care the project required.

Months after the first interview, the library hosted a small listening evening. Families sat beneath fluorescent lights while voices filled the room in two languages. The website was simple. Some transcripts still contained notes for further review. A hand-drawn sign explained how the material had been prepared and whom to contact with a correction or concern.

Layla's grandmother listened to her younger voice through an old recording and then to a child reading the translated story beside it. The machine hadn't understood the wedding, the rain, or the smell of the bread. It'd helped those memories cross a distance that time, language, and money had made difficult.

That's the promise of democratized intelligence when it remains grounded in responsibility. More people can begin. More people can contribute. More people can turn lived knowledge into something teachable, searchable, useful, or sustainable. Experts don't become less valuable; they become guides to a wider field of participants.

The limiting question is no longer only, "Do I have access?" It's also, **"What will I use this access to protect, build, and share?"**

AI can widen the doorway. We still decide whether the room beyond it becomes a marketplace of shortcuts or a workshop in which more people are able to do meaningful work.

Part V – AI in Everyday Life

"Science and technology revolutionize our lives, but memory, tradition and myth frame our response."

– Arthur Schlesinger

13

AI for Creativity

The first sound in Lena's studio each morning was the click of the kettle. The second was silence pretending to be helpful.

She'd been working on the same novel for three years. The story followed a mother returning to the town she'd left as a teenager, but every opening sounded as though it were trying too hard to announce what the book was about. Lena had written seven versions, deleted four, and printed three. They were taped above her desk with notes in the margins: too polished, too distant, starts before the real beginning.

The blank page wasn't empty. It was crowded by every book Lena admired, every sentence she'd abandoned, and the private fear that this time she'd reached the edge of what she could make.

She sat down anyway because sitting down was part of the work.

On that morning, instead of asking an AI assistant to write the chapter, she gave it a more useful role:

> ***"Give me ten possible ways to open a novel about a mother returning to a town that's chosen to misunderstand why she left. Don't write finished prose for me. Describe the opening situation, emotional distance, point of view, and central tension in two or three sentences each. Make the options meaningfully different, including two that may be uncomfortable or structurally unusual."***

Most of the ideas were wrong. One began with a funeral, which felt predictable. Another used a letter, which solved too much too quickly. A third placed the mother in a grocery store where everyone recognized her and no one spoke. That idea was also wrong, but it irritated Lena in a useful direction. The town wouldn't greet her with silence. It'd greet her with exaggerated warmth—the kind that makes clear a judgment has already been made.

Lena closed the response and wrote her own sentence.

It wasn't perfect, but it had a pulse. She read it aloud, changed one word, and felt the familiar warmth in her chest that meant a doorway had appeared.

Later, she might say the tool helped her find the beginning. The truth was more precise. It gave her something to resist. **The machine widened the hallway. Her voice still had to do the walking.**

That's the healthiest way to understand AI in creative work. It can help us begin, explore, vary, organize, and polish. It can offer more possibilities than fatigue or fear might allow us to imagine alone. But possibility isn't the same as art. Art begins when a person chooses what matters, accepts the cost of that choice, and shapes the result until it carries something only that person could've meant.

Open the Window

Every creative practice contains a moment when the field needs to widen. A novelist needs another way into a scene. A designer has stared at the same composition until every option looks inevitable. A musician can hear where a chorus should lift but not what chord will carry it. A teacher needs a visual metaphor for an idea students aren't yet feeling. A woodworker knows the function of a piece but not its shape.

AI is useful at this stage because it's willing to be wrong quickly.

That may sound like faint praise, but creative work often needs quantity before it can find quality. A model can generate alternate structures, unlikely combinations, thumbnail concepts, variations in tone, or questions the creator has stopped asking. The outputs aren't precious, which makes them easier to challenge. They create motion without demanding loyalty.

Lena began using the assistant as a room of movable walls. When a scene became too predictable, she asked how it might change if the person with the least power understood the most. When a character's motivation felt convenient, she asked for several explanations that'd make the character less sympathetic but more believable. When the middle of the book sagged, she asked the model to describe where tension was being released too early—not to rewrite the plot, but to help her see its shape.

Sometimes the suggestions were obvious. Sometimes they were beautifully phrased and completely wrong for the book. Occasionally, one exposed an assumption Lena had mistaken for a requirement. The value wasn't that the model knew the story. It was that the model could keep offering doors until Lena recognized the one she was afraid to open.

The same approach works across mediums. An image generator can explore lighting, palette, composition, and atmosphere before a painter or designer commits to a direction. A music tool can propose progressions or rhythmic variations that give a songwriter something new to hear. A video editor can prepare rough cuts, transcripts, captions, and alternate sequences so the creator can spend more attention on pace and meaning. A design assistant can produce several low-cost arrangements before materials are cut or a client becomes attached to the first idea.

Specificity improves this collaboration. "Give me something creative" invites the average of many things. A prompt that names the audience, constraint, emotion, material, and what must be avoided gives the system useful edges. It also forces the creator to articulate an intention before being distracted by attractive output.

Lena kept a small card beside the keyboard with five questions:

Who's this for? What should they feel? What must remain unresolved? What am I refusing? What would make the safer version less true?

Those questions mattered more than the name of the tool. They gave her a standard by which to judge what appeared.

Opening the window also means knowing when to close it. Endless variation can become a sophisticated form of avoidance. If another prompt always produces another option, a creator can remain in possibility forever and never risk choosing. Lena set a boundary: three rounds of exploration, then a direction. After that, the assistant could help her examine the choice, but it couldn't keep reopening every decision.

Abundance is useful only when it leads back to commitment.

The Human Work Is Choosing

The middle of creative work is less glamorous than the beginning. This is where the creator must decide what stays, remove what's merely clever, and continue long after novelty has faded.

Lena's first chapter improved, but it still sounded guarded. She could feel that the mother in the story was observing the town from behind glass. The prose was technically controlled and emotionally protected. Instead of asking the assistant to "make it more emotional"—an instruction likely to produce decorative intensity—she asked it to become a careful critic:

"Read this chapter as a demanding developmental editor.

Don't rewrite any sentences. Identify where the character appears to avoid the central emotional risk, where the reader may become confused, and where the prose explains what the scene has already shown. For each observation, quote the relevant phrase, explain the effect on the reader, and ask me one question that could help me revise it myself."

One observation bothered her. The assistant noted that the mother described every room in detail but never allowed herself to look directly at the person she'd returned to see. Lena's first reaction was defensive. Her second was recognition.

She printed the chapter and revised it by hand.

That sequence matters. AI can offer a critique, but the creator has to feel which criticism reaches the living center of the work. A model may notice repetition and miss a deliberate refrain. It may praise a line because it resembles successful writing elsewhere, while the author knows the line doesn't belong to this voice. It may recommend clarity where ambiguity is the point. The creator remains responsible for knowing the difference.

This is where taste becomes more valuable as generation becomes easier. Taste isn't simply preferring one color, phrase, or sound over another. It's the accumulated ability to recognize what serves the work. It grows through attention, practice, failure, and exposure to work made with care. AI can accelerate the number of options placed before taste. It can't live the years through which taste was formed.

Choosing also means accepting trade-offs. A quieter opening

may be less immediately dramatic but more faithful to the novel. An unusual design may confuse some viewers while making the right audience feel seen. A song may lose a catchy phrase because the phrase belongs emotionally to a different song. Every meaningful work is shaped as much by what the creator removes as by what the creator includes.

AI can support the maintenance surrounding those decisions. It can transcribe an interview, clean background noise, prepare captions, organize research notes, rename files, or turn a list of revisions into a production sequence. These tasks matter, but they can consume the attention needed for the work itself. Letting a system carry part of that practical weight can return an evening to the studio.

The recovered time should be visible. If automation merely creates room for more output, the creator may become faster and no less exhausted. Lena used the time to read the chapter aloud, walk without headphones, and return the next morning able to hear what the page was doing. The tool hadn't created the insight. It'd protected the conditions in which insight was more likely to appear.

Collaboration with other people remains essential. A model can simulate a skeptical reader, but it doesn't risk a friendship by telling the truth. It doesn't bring a history with the genre, a cultural experience the author doesn't share, or the embodied reaction of laughing, withdrawing, or becoming quiet at a particular line. Human readers offer more than feedback. They reveal how work lands in a life.

Lena used AI before the human read to remove avoidable confusion and organize the questions she wanted to ask. That allowed her readers to spend their attention on the deeper work: where they trusted the character, where they stopped believing, what stayed with them after the chapter ended, and what the book seemed unwilling to say.

The partnership worked because each participant had a different job. The machine gave breadth and tireless comparison. Human readers brought lived response. Lena carried the intention and made the final choice.

Protect the Part That's Yours

Faster beginnings and easier imitation create a responsibility that earlier tools didn't make quite so immediate. A person can now request work resembling a living artist's style, reproduce a voice, generate a face that never existed, or create a polished draft without understanding the sources from which its confidence was assembled.

Convenience can outrun conscience if the creator doesn't decide on boundaries before the tempting output appears.

Lena wrote a short creative covenant for herself:

> ***"Help me write a one-page creative-use policy for this project. Include what I may use AI for, what I won't outsource, how I'll handle private or copyrighted source material, when I'll disclose AI assistance, how I'll verify factual claims, and how I'll avoid imitating a living creator's***

distinctive style. Ask me questions before drafting any boundary you can't infer."

Her final policy was plain. She'd use AI for brainstorming, structural questions, research organization, and copyediting support. She wouldn't ask it to produce publishable scenes in another writer's voice. She wouldn't upload unpublished work belonging to someone else. She'd verify factual material against reliable sources. She'd keep records of substantial assistance and disclose it where readers, collaborators, clients, or publishers would reasonably expect to know.

The covenant didn't answer every question in the changing legal and creative landscape. It gave Lena a line she could recognize when speed made crossing it easy.

The deeper boundary concerned voice. Voice isn't a collection of sentence lengths or favorite words that can be extracted and reproduced. It's the pressure of a life against language: what a person notices, avoids, jokes about, mourns, and can't stop returning to. A model can imitate the surface of that pressure. It can't pay the cost that made the voice necessary.

That's why some creative acts should remain slow. A friend may deserve an apology written without a generated shield. A condolence may need the awkward honesty of a sentence that took ten minutes to form. A family story may need to retain the phrase that no editor would call elegant because it's the phrase the person actually used. Efficiency isn't neutral when slowness is part of the meaning.

There are also moments when assistance allows someone to

become more fully themselves. A person with limited mobility may use voice and image tools to express a design their hands can no longer draw cleanly. A parent may write during the ten quiet minutes before school pickup because an assistant helped recover the thread. A student who believed they "couldn't draw" may discover an eye for composition by directing, rejecting, and revising visual possibilities. A creator working in a second language may gain enough confidence in grammar to let the original idea take center stage.

The ethical question isn't whether a tool touched the work. Tools have always touched the work. The question is whether the process preserved consent, credit, truth, and the creator's real contribution.

As synthetic media becomes more convincing, audiences will care increasingly about origin. Clear disclosure isn't a confession that the work is less valuable. It's part of the trust between maker and audience. The most useful disclosure is specific enough to mean something: AI assisted with transcription and early structural questions; the author wrote and revised the final text. That tells the reader more than a vague label and avoids pretending the collaboration didn't occur.

Months after the morning with the ten openings, Lena stood in a small bookstore and read the first page of her finished novel aloud. The mother in the story entered town beneath a banner welcoming her home, and every person she passed smiled a little too brightly. The sentence that opened the book didn't resemble any of the suggestions the assistant had offered. It carried Lena's rhythm, her restraint, and the emotional risk she had almost avoided.

The tool had been present throughout the process. It'd offered alternatives, raised questions, organized notes, and caught inconsistencies. It hadn't decided what the novel believed.

That remained Lena's work.

Creativity in the AI era won't belong only to people who reject the tools or to people who use them for everything. It'll belong to people who understand the difference between assistance and authorship, between abundance and choice, and between an attractive output and a work they're willing to put their name on.

Open the window. Let more possibilities into the room. Then close it enough to hear your own voice—and stay long enough to make the work yours.

14

AI in the Home

At 5:30 on a Thursday evening, Daniel Park stood in the kitchen trying to decide what needed his attention first. Nothing was truly wrong, but the ordinary demands of family life had all arrived at once.

A carpool text arrived as the dog stared toward the back door with moral urgency. Nine-year-old Sophie needed poster board for a project she'd mentioned "a bunch of times." Her older brother, Miles, couldn't find the permission slip that'd been on the counter that morning. Grandma called from the pharmacy because the refill wasn't ready. Daniel Park opened the refrigerator, moved three containers as if dinner might be hiding behind them, and asked the question everyone had been avoiding.

"What are we eating?"

His wife, Erin, was still on a work call in the bedroom. She'd spent the week carrying the family calendar in her head—soccer, appointments, school deadlines, groceries, and the repair person who was supposed to arrive sometime between eight and noon. Daniel carried other parts of the load. Neither of them felt as though the work was shared evenly because most of the work was invisible until somebody forgot it.

Homes run on thousands of small decisions that disappear when they go well. Someone remembers the form, notices the milk is low, schedules the appointment, moves the laundry, changes the air filter, and knows which child has quietly outgrown a pair of shoes. No single task looks overwhelming. Together, they create a background question that follows people from room to room: **What am I forgetting?**

AI can't love a family. It can't repair a hard conversation, recognize the meaning of a long pause, or notice when a child needs a parent to stop solving and simply stay. But it can help carry part of the coordination that leaves people too depleted for those moments.

That night, dinner was cereal, eggs, and the last two pieces of toast. Nobody was harmed by it. After the children were in bed, Erin and Daniel sat at the table with the scattered calendars and agreed that the goal wasn't to make the home impressive. It was to make the home easier to live in.

Build a Home That Remembers

The Parks began on Sunday with one shared page and one rule: the system had to be simpler than the problem.

They brought together the commitments that couldn't

move—work hours, school dates, practices, appointments, and Grandma's transportation. Then they added the less visible work: groceries, forms, household maintenance, and the fifteen-minute tasks that became emergencies only because nobody remembered them until the last moment.

Erin asked an AI assistant to prepare a first version of the week:

> ***"Act as a family planning assistant. Using the schedules and notes below, create a realistic weekly plan with travel time, preparation time, and buffers. Surface conflicts instead of silently choosing between them. For each conflict, offer two options and explain the trade-off. Include a short list of items that must be prepared the night before. Don't schedule over protected family time or assume that every open hour is available."***

The plan was wrong in several places. It didn't know that Miles needed more transition time in the morning, that Erin avoided difficult calls immediately before school pickup, or that Grandma preferred an early appointment because waiting rooms became harder as the day went on. The family corrected those details together.

The first draft saved them from assembling the week from nothing. The conversation made the plan belong to the people who had to live it.

That distinction prevented the schedule from becoming another source of pressure. The assistant could propose an arrangement. It couldn't decide that efficiency mattered more than someone's energy, comfort, or need for margin. A home isn't a warehouse, and family members aren't resources to be allocated until every blank space disappears.

Once the weekly plan became visible, the Parks turned to remembering. They stopped relying on a single person to hold every promise. When someone said, "I'll take care of it," the task went to the place that person would actually see. A school form became a reminder with the document attached. A household repair went onto a shared list with a photo and the model number. Groceries entered one list rather than three text threads. Recurring maintenance appeared early enough to be ordinary.

The technology wasn't sophisticated. That was part of why it worked.

AI became useful when information was messy. It could turn a school newsletter into dates, forms, and questions. It could summarize a long email from the homeowners association and separate what required action from what was merely an announcement. It could help compare repair estimates, prepare questions for a contractor, or organize a list of concerns before a service appointment. The family still verified the details and contacted the right professionals. The assistant reduced the search and sorting around the decision.

Dinner improved in the same way. The Parks didn't ask for a perfect meal plan based on an imaginary family that enjoyed cooking every night. They gave the system the week they actually had: two late practices, a tighter grocery budget, Grandma's preferences, Sophie's suspicion of visible onions, and a Thursday when nobody should be trusted with a complicated recipe.

The result was a loose frame—two meals large enough for leftovers, one assemble-rather-than-cook night, one freezer backup, and one meal chosen for pleasure. A grocery app carried the shared list, and a person reviewed every order before it was placed. The benefit wasn't culinary optimization. It was that the family stopped spending its most tired hour asking the same question from the beginning.

Energy use offered another example of a home remembering its better choices. A thermostat schedule and a few monitored outlets helped the Parks notice patterns behind a higher bill. They shifted appliances away from peak times where that applied, reduced unnecessary standby use, and stopped cooling the house while a window was open. The changes were modest. Their value came from being repeated without requiring a fresh decision each day.

A useful home system works like that. It carries the routine until the routine becomes quiet.

Design for Real People

The first week looked orderly on the shared page and felt impossible to Miles.

He was thirteen, funny, bright, and able to lose track of time while standing beside a clock. A reminder that said "Leave for school" arrived too late because leaving contained several tasks: stop what he was doing, find his shoes, pack the charger, locate the water bottle, and tolerate the feeling of being rushed while everyone watched.

Erin had spent years repeating instructions with increasing volume and decreasing hope. The new system gave the family a chance to design the transition instead of judging the person struggling with it.

They asked for a more useful routine:

"Create a respectful fifteen-minute leaving-the-house routine for a teenager who struggles with time awareness and transitions. Use no more than five visible steps, include a five-minute and two-minute cue, and suggest how the materials can be staged near the door. Avoid shame, childish language, constant tracking, or rewards that turn every morning into a negotiation."

The family tested the suggestions and kept only what helped. A visual timer made time easier to feel. A small landing place near the door reduced the search for daily items. The first cue invited Miles to finish what he was doing rather than demanding an instant switch. Some mornings still fell apart. The difference was that difficulty no longer became evidence of bad character.

Good household technology supports a person without turning that person into a problem to be managed.

Grandma's needs made the same lesson visible in another direction. After a fall, the family wanted more reassurance. It'd have been easy to fill the house with sensors and call the result safety. Grandma wanted support, but she didn't want every movement converted into a report.

They began by asking her what would make the house easier rather than what would make the family less anxious. She chose larger text on the shared tablet, captions by default, better lighting near the stairs, medication reminders she controlled, and a simple way to call for help. She declined a camera at the front door and constant location tracking. The family respected the answer.

Before medical appointments, an assistant helped organize questions and summarize information the family had already verified. It didn't diagnose, interpret a concerning change as harmless, or recommend altering treatment. The purpose was to arrive prepared enough that Grandma could use the appointment to discuss what mattered rather than spend half of it reconstructing dates.

Homes differ, so the design must differ too. A single parent with a changing shift schedule may need help identifying collision days and drafting a carpool request that still sounds personal. Co-parents may need one neutral place for dates, school notes, and agreed decisions so old frustration doesn't have to travel inside every handoff. Roommates may need a shared page for bills, chores, and staples before polite resentment becomes the household communication system. A person living alone may want reminders and maintenance support without inviting more surveillance into the space.

The principle remains the same: begin with the people, not the features.

Ask what makes mornings hard, where information disappears, which decisions repeat, and what kind of help would feel respectful. Invite everyone affected by a routine to shape it. Consent inside a home shouldn't be a one-time click by the person who bought the device. It should be an ongoing conversation about what's collected, who can see it, and whether the benefit is worth the intimacy it requires.

The Parks also kept a paper binder near the router. It held emergency contacts, medication information, insurance details, instructions for shutting off the water, and the few phone numbers nobody remembered because they lived in devices. When a storm interrupted power and internet service, the home remained understandable.

Paper still works in the dark.

Keep the House Human

As the household system improved, Erin noticed a temptation to keep going. If reminders helped with forms, perhaps the family could track chores more closely. If energy data reduced a bill, perhaps more sensors would reveal more savings. If schedules prevented conflict, perhaps every hour should've a plan.

This is where a useful system can begin to change the atmosphere of a home. The dashboard becomes more visible than the people. Every deviation looks like failure. The family starts managing itself as though life were a performance problem.

The Parks wrote a boundary at the top of the shared page: **Automate coordination, not connection.**

Bills could be paid automatically. Groceries could live on a shared list. A school email could become a reminder. But an apology would be spoken by the person who owed it. The system wouldn't decide when someone needed a hug, send affection on a schedule, or interrupt a difficult story because the calendar said it was time to move on.

Dinner remained screen-free. Not every meal was meaningful, and not every conversation sparkled. That wasn't the standard. The value came from leaving enough quiet for something unscheduled to happen: a joke that lasted too long, a question that revealed worry, or Grandma telling the children why the smell of a particular soup reminded her of home.

Privacy received the same deliberate attention. The family reviewed which devices listened, watched, stored, or shared. They disabled features they didn't use, preferred local processing where practical, removed old accounts, and avoided placing cameras in private spaces. The children knew what adults could see. Nobody used hidden monitoring to create obedience.

Once a month, they ran a short home-technology review:

> ***"Using the household tools and routines listed below, identify what each one collects, what purpose it serves, who's affected, and what happens if it fails. Flag unnecessary tracking, duplicated tools, unclear permissions, and any routine that saves time by reducing someone's digni-***

ty or choice. Recommend what to keep, simplify, turn off, or replace with a non-digital option."

The review often removed more than it added. A neglected app disappeared. Two reminders became one. A sensor whose data nobody used was disconnected. The family learned that simplicity wasn't a lack of sophistication. It was a way to keep the system legible enough that everyone could understand the bargain being made.

Attention also needed protection from convenience. A voice assistant that can answer every question may keep a child from wondering long enough to investigate. A recommendation system can fill an evening so completely that boredom—the old doorway to imagination—never arrives. The Parks didn't reject the tools. They left some spaces unfilled. A board game didn't need optimization. A walk didn't need a score. The backyard could remain a place where the children noticed what grew there without an app naming every leaf.

Several Thursdays after the cereal dinner, the house was busy again. The dog still needed to go out at the worst possible moment. Sophie remembered an assignment later than anyone would've liked. Grandma's appointment moved. No system had made the family frictionless.

But the week was visible. Dinner had been chosen before hunger made the choice feel personal. Miles's bag waited near the door. The appointment change appeared in one place. Erin and Daniel weren't each carrying a different version of the family in their heads.

At the table, Sophie described a problem with a friend. The story wandered. Nobody rushed her.

That was the return the system had been built to create—not perfect logistics, but enough recovered patience for a small person's complicated story to receive the room it needed.

The best AI-enabled home won't feel like a showroom. It'll feel like a place that remembers what can be remembered, adapts to the people inside it, and becomes quiet when human life asks for the foreground.

15

AI for Lifelong Learning and Hobbies

The dishes were done, the house had finally exhaled, and Ravi's guitar waited in the corner like a friend he kept promising to call.

He'd bought it eight years earlier with the sincere belief that ownership would eventually become ability. Every few months, he began again. He watched a long beginner video, tried to remember four chords, and stopped when his fingers hurt and the sound coming from the instrument bore little resemblance to music. The guitar returned to the corner, where it became less of an invitation and more of a quiet accusation.

Ravi didn't lack interest. He lacked a path small enough to survive the rest of his life.

That evening, his daughter mentioned that she wanted everyone to perform something at her birthday dinner. She was joking. Ravi looked at the guitar and decided that one imperfect song would be a better gift than another year of waiting to become the kind of person who played.

He opened an AI assistant and described the real situation: forty-four years old, complete beginner, twelve minutes available after dinner, sore fingertips after long sessions, and six weeks before the birthday. He named one simple song his daughter loved and asked for a plan that'd make a recognizable verse possible without pretending he could become a guitarist in forty-two days.

> ***"Design a six-week beginner guitar plan for someone with twelve minutes after dinner each day. The goal is to play one recognizable verse of the song below at a family birthday, not to master the instrument. Build each session around tuning, one small technique, one chord-change drill, and time with the actual song. Include a weekly recording checkpoint, protect against overuse or pain, and adjust the plan when my notes show that a skill needs more time."***

The response was modest: tune, warm up, practice two chord changes, play a slowed section of the song, and record twenty seconds at the end of the week. The plan didn't inspire Ravi with a vision of mastery. It did something better. It told him what to do tomorrow.

AI's most useful role in lifelong learning isn't to make learning effortless. Learning requires attention, repetition, correction, and the humility of being bad at something long enough to become less bad. The tool can lower the friction around that process. It can turn a large ambition into a sequence, explain an idea another way, prepare practice material, and help a person notice the next correction.

It can't place Ravi's fingers on the strings. It can't develop his ear while he avoids listening. It can't feel the moment a chord finally rings cleanly because his hand has learned what his mind couldn't fully explain.

AI can carry the scaffolding. The learner still has to climb.

The Loop, Lived

Ravi's first week was almost comically small. He learned how to hold the pick, formed two chords, and discovered that the distance between them felt much larger to his hand than it looked on a diagram. The assistant had suggested moving slowly enough to keep the sound clean, but Ravi kept speeding up because speed felt like progress.

On the fourth night, he recorded thirty seconds and asked for help diagnosing the problem:

> ***"I'm a beginner practicing the chord change shown in this recording. Give me one correction I can test in the next five minutes. Explain what you noticed, how I can tell whether the adjustment is helping, and one common mistake to avoid. Don't give me a full lesson or more than one change***

at a time."

The feedback pointed to his wrist angle and the finger muting a neighboring string. Ravi tried the adjustment, recorded again, and heard a cleaner chord. The improvement was small enough that nobody else in the house would've noticed. He noticed.

That's the learning loop in its most useful form: **plan, practice, receive feedback, reflect, and repeat**.

AI can support every part of the loop. It can build a plan around the available time rather than an ideal schedule. It can generate a fresh practice problem when a learner has memorized the answer instead of understanding the principle. It can explain the same concept with a diagram, analogy, example, or question. It can compare two attempts and help identify a pattern. It can turn a week of notes into a short reflection on what's improving and where the next bottleneck may be.

The quality of the loop matters more than the sophistication of the tool. A learner who receives fifty suggestions at once hasn't received feedback; they've received another source of overwhelm. Ravi improved faster when he asked for one correction, tested it, and listened. The tool became most useful when it respected the narrow bandwidth of actual practice.

This same rhythm can support almost any skill. A language learner can rehearse a five-minute conversation, receive gentle corrections, and try it again with different vocabulary. A painter can share a study and ask for one observation about value or composition, then decide whether the critique fits the intention. A chess player can examine the moment before the visible mistake and ask what question should've been asked. A home cook can choose one cuisine for a month, learn a few core techniques, and build meals around shared ingredients instead of collecting hundreds of recipes.

The tool shouldn't complete the learning task that the learner needs to practice. If the purpose is to learn to write, generating the final essay removes the repetition through which writing improves. If the purpose is to understand algebra, accepting the answer without working the steps creates fluency in submission, not mathematics. If the purpose is to develop visual judgment, scrolling through generated images without making choices exercises preference more than craft.

The useful boundary is simple: **offload the friction that blocks practice; keep the effort that creates the skill**.

For Ravi, tuning with guidance reduced friction. A short plan reduced friction. Slowing a song and looping four measures reduced friction. Forming the chord, hearing the buzz, changing his hand, and trying again created the skill. The first group made the second group easier to reach. It didn't replace it.

As the weeks passed, the assistant adjusted the plan because Ravi's progress was uneven. One chord change improved quickly; another remained stubborn. A rigid course would've moved on. A responsive tutor could give him a different exercise and a more useful explanation. Ravi still made the choice to stay with the difficult change rather than hide inside the parts he already performed well.

That willingness to remain with the hard part is one of the qualities no tool can automate on a learner's behalf.

Build a Cockpit You Can Return To

Many learning plans fail outside the lesson. The learner knows what to do and still doesn't return often enough for the skill to take root.

Ravi's breakthrough wasn't a more powerful app. It was moving the guitar.

He placed it on a stand beside the table where the family ate, kept the pick in a small dish, and attached practice to a cue that already existed: when the dinner dishes were finished, he played for twelve minutes before sitting down with his phone. The instrument became easier to reach than the distraction.

A learning practice that survives real life needs only a small cockpit: **one visible plan, one reliable cue, one place to practice, and enough evidence of progress to know where to return.**

The plan should fit the season. A parent with young children may have ten minutes in a pickup line. A retired person may have longer blocks but need help choosing a direction. Someone recovering from an injury may need a teacher or clinician to define safe limits. A person learning for career growth may need projects that create evidence of skill, while someone learning for joy may need protection from turning every interest into another performance review.

Ravi's cockpit lived on one page. It showed the song, the chord sequence, the current tempo, one weekly focus, and links to three recordings. There was no elaborate dashboard. The page took less time to maintain than the practice itself.

On Sundays, he used a short review prompt:

"Using the practice notes below, help me review the week. Identify one thing that improved, one recurring obstacle, and one adjustment for next week. Keep the sessions within twelve minutes, preserve time for playing the actual song, and ask whether the plan still feels enjoyable. Don't add a new tool unless the current process can't solve the problem."

The final instruction protected him from a common trap. Learning applications can make setup feel productive. Courses, playlists, flash cards, note systems, and progress trackers multiply until the learner spends more time arranging the cockpit than flying.

Choose the smallest stack that supports the loop. A general assistant can help plan and explain. A specialized tool may provide the feedback unique to the skill: pronunciation, pitch, notation, spaced repetition, coding tests, or game analysis. A calendar or timer protects the return. A notebook, folder, or simple page holds the evidence. Everything else has to justify the attention it requires.

The cockpit should also preserve context. When a person returns after a difficult week, they shouldn't have to reconstruct the entire project. The page can say: You were working on this chord. Your last recording showed this improvement. Begin here. That small act of remembering prevents interruption from becoming abandonment.

Projects give the learning a reason to continue. Ravi wasn't practicing chord changes in the abstract. He was preparing a birthday verse for someone he loved. A language learner might plan a voice message to a grandchild. A beginner coder might automate one repetitive task for a community group. A woodworker might build a small shelf for a particular wall. A gardener might create a pollinator bed and document what survives the season.

The project should be small enough to finish and real enough to matter.

Completion changes identity. A person who finishes one wobbly song has evidence that they can return to an instrument. A person who holds a short conversation has crossed from studying a language to using it. A learner who builds a simple working script may not be a professional developer, but they've become someone capable of making a computer carry one useful piece of work.

That evidence is more motivating than an abstract streak because it connects practice to a life.

Protect the Joy and the Skill

By the fifth week, Ravi could play the verse slowly. That was when he nearly ruined the project by becoming ambitious.

He began comparing himself with musicians online. Twelve minutes became thirty. He added a second song, downloaded another course, and spent an evening researching equipment that wouldn't improve the chord he still couldn't play cleanly. The original promise—to make one small gift—started disappearing beneath a new demand to prove that the effort counted.

AI can intensify this problem because it makes expansion effortless. A learner can always request a larger plan, another project, a faster path, or a way to monetize the new skill. Curiosity becomes content. A hobby becomes a side business before it's had time to become a refuge.

Ravi returned to the reason he began. He removed the second song, restored the twelve-minute limit, and stopped recording every session. Some practice remained private. Some mistakes were allowed to disappear without becoming data.

Not every hobby needs to become income, a public identity, or proof of discipline. People need forms of learning that are allowed to be useless in the economic sense and deeply useful in the human one. Painting can be a way to see. Gardening can be a way to live inside a season. Music can be a way to remain near beauty without needing to become impressive.

Protecting the skill also means knowing when AI isn't the right teacher. Physical crafts can involve tools, heat, weight, chemicals, repetitive strain, or techniques that require hands-on correction. Medical, athletic, and safety-related activities need qualified guidance appropriate to the person and situation. A confident generated instruction doesn't become safe because it's detailed. Use AI to prepare questions, organize reputable material, and support practice within boundaries established by people who understand the risk.

Children's learning deserves its own care. A tutor that adjusts patiently can be valuable, but a child should understand when they're interacting with a system, what information it keeps, and when to ask an adult. The goal isn't to create a perfectly measured learner. It's to help a young person develop curiosity, confidence, and the ability to struggle without believing that struggle means they're incapable.

Adults need the same grace. Progress is rarely linear. Work becomes busy. A family member needs care. Energy changes. The skill goes untouched for two weeks and begins to feel farther away than it is. A well-designed system doesn't shame the learner back into motion. It remembers where they stopped and offers a smaller way to return.

On the evening of his daughter's birthday, Ravi stood in the living room with the guitar strap twisted and his fingers colder than they'd ever felt during practice. The first chord buzzed. He paused, laughed, and began again.

The verse was slow. One change arrived late. His daughter recognized the song before anyone else did and started singing with him.

Nobody in the room mistook the performance for mastery. That wasn't what made it successful. The song carried six weeks of small returns: the dish put away, the guitar lifted, the difficult chord attempted once more. AI had helped Ravi build the plan, diagnose a few problems, and remember where to begin each night. It hadn't played the song or given the song its meaning.

The meaning came from the person who practiced and the person for whom he practiced.

That's the promise of AI in lifelong learning. It can make the next step clearer, the feedback more available, and the path easier to reenter. It can help turn someday into twelve minutes tonight.

Then the learner has to show up.

Part VI – Living Beyond Human Limits

"Twenty years from now you will be more disappointed by the things that you didn't do than by the ones you did do. So throw off the bowlines. Sail away from the safe harbor. Catch the trade winds in your sails. Explore. Dream. Discover."

– Mark Twain

16

Designing Your AI-Integrated Life

On an ordinary Tuesday, Samantha's phone began humming before her eyes had fully opened.

A calendar reminder appeared for a meeting she didn't remember accepting. An email thread had copied her for "visibility." A fitness app congratulated her for yesterday and warned her about today. A grocery notification suggested that she reorder something the family no longer used. Before Samantha's feet reached the floor, five systems had already expressed an opinion about what should happen next.

She answered two messages in bed, checked the calendar, and felt the day pull away from her.

Nothing about the morning seemed dramatic. That was precisely the problem. Technology rarely announces that it's rearranging a life. It enters as a convenience, becomes a routine, and eventually begins shaping attention by default. AI can accelerate that drift because it can draft, schedule, recommend, summarize, and nudge before a person has decided what those actions should serve.

By 9:15, Samantha was in her fourth thread and hadn't begun the work she'd called most important the night before. She wasn't disorganized. She was highly organized around other people's access to her.

The question wasn't whether she should use less technology or more of it. The question was whether the system surrounding her reflected the life she meant to live.

That evening, Samantha wrote two sentences on a sheet of paper.

I want technology to reduce maintenance, not increase urgency.

I want the time it returns to become visible in my health, my work, and the people I love.

Those sentences became the beginning of her personal operating system—not a complicated dashboard or a collection of automations, but a set of choices about what machines could carry, what remained hers, and how she'd know when convenience had begun to take more than it gave.

Start with Intent, Not Tools

The easiest way to build an AI-integrated life is to begin with applications. The healthier way is to begin with friction and meaning.

Samantha listed the parts of the week that repeatedly drained her. At work, status requests and scattered notes consumed the hours when she thought most clearly. At home, the family reached dinner with no plan and very little patience. She postponed financial reviews because the accounts felt disconnected. Exercise followed an exhausting pattern of doing too much and then doing nothing. A notebook full of ideas sat beside the bed, but creativity received whatever attention remained after the phone was finished with her.

Then she made a second list: the parts she didn't want to optimize away.

She wanted to write the opening of important messages herself. She wanted family decisions to remain conversations. She wanted exercise to include the feeling of being in her body, not only the score assigned afterward. She wanted prayer and quiet reflection to remain outside any system that turned them into streaks. She wanted her children to experience her face, not the top of her phone.

The two lists created a boundary. The first showed where assistance might help. The second showed what the assistance was for.

Samantha used an AI assistant to organize the reflection, but not to choose the values:

"Help me audit the week described below. Identify re-

peated friction, decisions I keep remaking, information I repeatedly search for, and tasks that consume attention without requiring my judgment. Then identify moments where presence, privacy, creativity, relationship, or accountability may be more important than efficiency. Suggest three small experiments, not a complete life overhaul, and explain what each experiment should protect."

The response proposed more changes than Samantha needed. She chose three.

First, her inbox would no longer receive the best hour of her day. An approved assistant could separate likely decisions from routine updates and prepare a short morning brief. Samantha would review the source messages before acting, especially when tone, politics, or incomplete context mattered.

Second, the family would spend fifteen minutes on Sunday making the coming week visible. AI could draft the schedule and meal options from the information they provided. The family would decide what was realistic and what needed to remain flexible.

Third, her phone would charge outside the bedroom. The paper beside the bed would hold the ideas that arrived at night. The world could wait until she'd stood inside her own morning for a few minutes.

These weren't dramatic changes. That made them more likely to survive.

Intent also protected Samantha from the fantasy of the perfect plan. AI can generate a beautiful schedule that ignores traffic, fatigue, a child's difficult day, or the fact that some work takes longer because it matters. A life isn't a clean dataset. The plan must bend without turning every change into evidence that the person has failed.

Samantha learned to judge a system by a human standard: **Does this make the life easier to inhabit?**

A calendar that protects focus but leaves no room for lunch isn't intelligent. A health plan that produces anxiety every time a metric changes isn't supportive. An automation that sends a technically correct message and damages trust hasn't saved time. The output may be efficient while the whole arrangement is unwise.

Some parts of life deserve to remain deliberately unoptimized. Dinner can last longer than planned. A hobby can produce nothing sellable. A walk can wander. A conversation can have no action item. These aren't defects in the operating system. They're reminders that the system serves a person rather than the other way around.

Build Your Personal Operating System

After two weeks, Samantha's three experiments had created enough relief to reveal the next question: how should the pieces work together without becoming another project to manage?

She built the operating system around four simple movements: **notice, decide, support, and review**.

Notice meant giving useful information a dependable place to land. Meeting decisions went into one shared workspace instead of remaining in transcripts, inboxes, and memory. Household commitments entered one calendar. Ideas went into one note rather than six open tabs. Financial information remained in secure tools designed for it. The goal wasn't to capture everything. It was to stop repeatedly losing what she'd already decided mattered.

Decide meant separating a recommendation from a commitment. AI could rank messages, compare options, or propose a schedule. Samantha still chose what deserved time, which trade-off was acceptable, and who might be affected. The system reduced the work around the decision without disguising the fact that a person was making it.

Support meant allowing routine help to continue after the decision. A focus block could be defended on the calendar. A meeting summary could carry owners and dates to the right place. A grocery list could remain synchronized. A recurring transfer could support a financial intention. A learning reminder could bring her back to the lesson she'd already chosen. This was Samantha's **Automation Floor**—the baseline support she no longer needed to rebuild each week.

Above that floor sat the **Human Ceiling**: strategy, judgment, care, ethics, taste, and the moments in which being present was the work. Samantha protected those moments from automation even when a tool could imitate part of them. She could ask for help structuring a difficult conversation, but she'd have the conversation. She could request a first analysis of a problem, but she'd defend the recommendation. She could prepare a family plan, but she'd listen when someone said the plan didn't work for them.

Review kept the floor from rising until it touched the ceiling.

Every Sunday, Samantha spent twenty minutes with a short prompt and the week's notes:

> ***"Prepare a weekly review from the information below. Summarize what required my best judgment, what created avoidable friction, what the current automations handled well, and where a tool created confusion or extra work. Suggest one routine to keep, one to adjust, and one to remove. Then ask me where I felt most present and where I lost ownership of the week."***

The last two questions prevented the review from becoming a productivity scorecard. A week could contain fewer focused hours because a child was sick and still reflect the right priorities. A system that couldn't recognize that difference would eventually train Samantha to mistrust her own humanity.

At work, the operating system changed her mornings. A brief showed the few messages that required decisions, with links to the originals. The calendar protected a ninety-minute block for strategy before meetings began. Meeting notes captured decisions, not every spoken sentence, and moved actions into the shared workspace. Samantha still wrote the first line of a delicate message and reviewed anything carrying her authority.

The time returned gradually. At first, it was twenty minutes. Then a meeting disappeared because the shared brief answered the question. Then a recurring report became a live view with a short narrative explaining what had changed. Samantha was no longer spending her leadership around the edges of leadership.

At home, the operating system was less formal. Sunday planning made the collisions visible before they became arguments. A short meal frame reduced the most exhausted decision of the day. School messages became dates and questions, but a parent reviewed them before anything reached the calendar. The family didn't automate warmth. It reduced some of the background work that'd been crowding warmth out.

Health required restraint. Samantha used a wearable to notice broad patterns in sleep and recovery, not to diagnose herself or let one score decide what her body was allowed to do. When a trend concerned her, she treated it as information to discuss with a qualified professional. The practical benefit was permission to respond earlier to fatigue rather than waiting until exhaustion turned into guilt.

Creativity returned in a quieter way. The notebook beside the bed began to fill. Once a week, Samantha chose one idea and asked an assistant to help her explore questions, possible structures, and what she might be assuming. Then she closed the tool and wrote for thirty minutes. She didn't need to become a professional artist. She needed a part of the week in which usefulness wasn't the only standard.

The operating system worked because the tools weren't the

center. They were a supporting layer around choices Samantha had made on purpose.

Protect the Life the System Serves

Several months into the experiment, the system began working so smoothly that Samantha almost stopped noticing it. That was the goal—and a reason to remain alert.

The deepest risk of personal AI may not be obvious failure. It may be a system that works well enough to become invisible while gradually choosing more of the life. If it decides what you read, when you rest, how you reply, what you eat, which route you take, and which relationships receive attention, convenience can become a form of quiet handling.

A recommendation becomes a default. A default becomes a habit. A habit begins to feel like identity.

Samantha protected her agency with three boundaries.

The first was **visible authority**. Every tool needed a clear role and a point where its authority ended. The email assistant could prioritize; it couldn't send sensitive responses. The calendar could propose; it couldn't fill protected time without permission. The family planner could organize; it couldn't decide whose needs mattered most. If a system made an important choice, Samantha wanted to be able to see that choice and reverse it.

The second was **data restraint**. Samantha reviewed what each tool collected, why it needed the information, how long the information remained, and whether a less invasive option could achieve the same purpose. She kept sensitive work inside approved environments and avoided placing medical, financial, family, or confidential information into general systems without appropriate safeguards. Convenience didn't automatically deserve intimacy.

The third was **human practice**. Samantha intentionally kept a few calluses. She wrote some pages without assistance. She performed mental calculations when understanding mattered. She found her way without navigation on familiar streets. She sat in difficult conversations without a generated script open beside her. These practices weren't a rejection of technology. They kept essential capacities available.

Once a month, she used a broader audit:

"Review the personal system below for dependence, privacy, and alignment. Identify any tool that's making consequential choices invisibly, collecting more information than its purpose requires, weakening a skill I want to keep, or consuming more attention than it returns. Suggest where I should add approval, reduce data, restore a manual practice, or remove the tool entirely."

The audit sometimes produced an uncomfortable answer. A news summary that claimed to save time had become another feed. A health metric was affecting Samantha's mood more than her choices. An automation created so many exceptions that maintaining it took longer than the task it replaced. Removing those tools wasn't failure. It was evidence that the review loop was working.

She also protected periods in which the system had no claim on her attention. The phone stayed downstairs at night. Dinner remained mostly screen-free. One block of the weekend had no productivity target. Whether she called it a digital sabbath or simply time off didn't matter. The practice reminded her that attention could still exist without being captured, measured, or improved.

The future will likely make this boundary harder. Personal assistants may learn the rhythms of a person's work, home, health, and relationships well enough to coordinate across them. They may notice patterns a person misses and provide genuinely valuable support. They may also create deep dependence, intimate data trails, and recommendations shaped by incentives the user can't see.

No fixed rule will resolve every trade-off. The durable question is the one Samantha wrote at the beginning: What's this helping me protect?

On another Tuesday, months after the morning that began in bed with two messages, Samantha woke before the phone did. It was charging downstairs. She wrote one idea in the notebook, dressed, and entered the kitchen.

A quiet brief showed three decisions that required her attention after breakfast. The calendar protected the first part of the workday. A family conflict had been surfaced with two options, neither chosen. The system had prepared the day without claiming it.

Later, an unexpected problem changed the plan. Samantha moved the focus block, canceled something that no longer mattered, and spent an hour with a person who needed her. The operating system didn't break. It bent around the reason it existed.

That's the goal of an AI-integrated life. Not a life optimized until every hour is productive, but a life in which routine friction occupies less of the mind and intention occupies more of it.

Use machines to remember, organize, compare, and prepare. Keep for yourself the authority to decide what the day is for.

The goal isn't to become more machine-like. It's to use machines in ways that leave you more deliberate, expressive, connected, and alive.

17

The Ethics of AI and Human Choice

The letter arrived in Evelyn's queue on a Monday morning.

"I understand that my application was declined," it began. "What I don't understand is why."

The writer, Nia, had applied for a small-business loan. She'd provided tax records, bank statements, and the other information requested. The response she received contained a score, a general list of possible factors, and no explanation specific enough to help her understand the decision or challenge a mistake. Somewhere inside the lender's process, a model had weighed patterns she couldn't see and converted them into an answer that affected what she might be able to build.

Evelyn led the team responsible for the decision-support system. She knew the model had performed well in testing. It processed applications consistently, surfaced risk quickly, and reduced the time employees spent reviewing routine cases. On the dashboard, the system looked like a success.

Nia's letter made the gap between performance and responsibility impossible to ignore.

A playlist can afford to be mysterious. A recommendation about a movie can be wrong without changing the direction of a life. A system involved in credit, employment, housing, health, education, insurance, or public benefits operates under a different moral weight. When an automated judgment shapes a person's opportunity, the person deserves more than an efficient answer. They deserve to understand what happened, correct what's wrong, and reach someone who has the authority to reconsider the result.

Evelyn printed the letter and wrote a question at the top: **What does this system feel like from the other side?**

That's where AI ethics becomes real. It's easy to discuss fairness, transparency, privacy, and accountability as abstract principles. The principles become clearer when we imagine the person receiving the output, the employee asked to trust it, and the family living with a mistake.

Artificial intelligence doesn't remove human choice from a system. It can make the choice harder to see. People still select the goal, the data, the threshold, the permissions, the explanation, and the point at which a human may intervene. Ethics is the work of making those choices visible before scale turns them into consequences.

Five Questions Beneath Every System

Evelyn began the review with five questions. They weren't questions only for engineers. They belonged to anyone approving, purchasing, using, or being affected by an AI system.

The first was **fairness: who's helped, who's burdened, and whose history is missing?**

The model had learned from previous lending data. That sounded objective until the team examined what the history contained. Past decisions reflected which businesses had been encouraged to apply, whose income fit traditional categories, which communities had easier access to banking, and which applicants had enough time or confidence to challenge an error. Historical data didn't merely describe the past. It carried the shape of past opportunity into the model.

Fairness didn't require every applicant to receive the same answer. It required the team to test whether similar situations were being treated consistently, whether error rates differed across relevant groups, and whether a variable was standing in for something the organization shouldn't use. It also required attention to the people missing from the data. A model can appear accurate when the people it serves poorly are too few or too quiet to disturb the average.

The second question was **transparency: can a person understand the reason well enough to respond?**

The team's original explanation had been designed to satisfy a process, not to support a human being. It listed broad possibilities because broad language was easier to generate. Nia couldn't tell which information mattered, whether a record was wrong, or what evidence might change the result.

Transparency doesn't mean exposing every line of code or overwhelming a person with technical detail. It means offering an explanation at the level of the decision: what information was important, what limitation affected the outcome, what remains uncertain, and what the person can do next. If an organization can't explain a consequential decision in language the affected person can use, it hasn't finished designing the decision.

The third question was **privacy: what information are we collecting, and did the person meaningfully agree to that use?**

AI grows more capable when it can connect more signals. That creates a temptation to treat every available detail as useful: purchases, location, browsing, communication patterns, or inferences about a person's behavior. But information gathered for one purpose doesn't automatically belong in another. A person who shared data to receive a service may not reasonably expect it to influence an unrelated judgment.

The team reviewed every category of information and asked whether it was necessary, whether a less invasive alternative existed, how long the data remained, who could access it, and what would happen if it were exposed. "We might use it someday" wasn't a purpose. Collecting less reduced both ethical risk and the amount of trust the organization had to ask customers to lend.

The fourth question was **autonomy: is the system assisting a**

choice or quietly taking the choice over?

Employees had begun treating the model's score as a conclusion even though the policy described it as decision support. The interface placed the score at the top, highlighted it in color, and made the underlying evidence harder to reach. Technically, a human made the final decision. In practice, the design taught the human to agree.

Evelyn's team changed the workflow so reviewers saw the important evidence, uncertainties, and reasons before the recommendation. They were asked to record their own judgment and explain disagreements in either direction. Human review became meaningful rather than ceremonial.

The fifth question was **accountability: who has the authority and obligation to act when the system is wrong?**

"The AI decided" isn't an answer. Engineers choose data and objectives. Product leaders choose the interface. Executives choose the level of risk the organization accepts. Operators choose how the recommendation enters the workflow. Legal and domain experts interpret the obligations that apply. Accountability requires named owners who can pause the system, investigate a failure, communicate with affected people, and make repair possible.

Evelyn turned the five questions into a first-pass impact review:

> ***"Evaluate the proposed AI use described below through five lenses: fairness, transparency, privacy, autonomy, and accountability. Identify the people affected, the ben-***

efits promised, the most plausible harms, the data required, the explanation a person would receive, the human override, and the named owner if something goes wrong. Treat missing information as a question, not permission to assume."

The prompt didn't make the system ethical. It made it harder for the team to hide from the questions.

Dignity Is the Design Test

The five questions pointed Evelyn toward a deeper standard: dignity.

Dignity asks whether the person remains visible inside the system. Does the design recognize that an applicant is more than a risk score, an employee more than a productivity metric, a child more than a learning profile, and a patient more than a pattern in a record? Does the person have a voice, a boundary, and a path to challenge the conclusion?

This test matters because AI can make a harmful arrangement look orderly. A workplace may introduce monitoring software in the name of efficiency and begin treating every pause as lost output. A school may adopt an adaptive tutor that helps students learn while quietly collecting a detailed record of hesitation, mistakes, and behavior. A voice assistant may feel deeply personal while its recommendations are shaped by commercial incentives the user can't see. In each case, the capability may be useful. The ethical question is whether the use preserves agency and trust.

Evelyn saw the workplace version in her own review team. Managers wanted to measure how quickly employees processed applications after the new system arrived. Speed was easy to count, so it threatened to become the definition of performance. But careful reviewers often took longer because they investigated missing information, recognized unusual circumstances, or spoke with an applicant before deciding.

If the organization rewarded only speed, it'd train people to behave like the narrowest version of the machine.

The team changed the measures. Cycle time still mattered, but it sat beside quality, reversals, customer understanding, appropriate escalation, and whether reviewers identified problems the system missed. The purpose of automation was to remove routine burden so people could apply better judgment—not to make human work imitate automated throughput.

Children and vulnerable people require an even higher standard. An AI tutor can offer patient explanation and adapt practice to a learner's pace. It shouldn't use a child's uncertainty as an advertising profile, make essential support dependent on invasive collection, or create consequences the child can't understand or appeal. Consent must be written for the person giving it, not for the lawyers defending it. Breaks should be possible. A parent, teacher, clinician, or other responsible adult needs a real override when the system crosses its role.

Synthetic media creates a different dignity problem: it can borrow a person's face, voice, or authority without permission. The same techniques that support dubbing, accessibility, education, and art can also fabricate evidence and make a lie feel remembered. Ethical use requires clear consent, specific disclosure, and ways to verify origin. A label shouldn't be hidden where only the determined reader will find it. The person encountering the media should be able to understand what was created or altered before trust is demanded.

Environmental cost belongs in the same conversation. AI systems depend on physical infrastructure, energy, water, hardware, and supply chains. Responsible design matches the scale of the tool to the importance of the task rather than using the largest available system by habit. Efficiency, reuse, and provider transparency matter because intelligence doesn't become weightless merely because the interface feels invisible.

Evelyn asked the team to rewrite the purpose of the lending system without mentioning technology. The first attempt said, "Improve decision efficiency and risk consistency." It was accurate and incomplete.

The final version said, **"Help qualified reviewers reach timely, evidence-based lending decisions while giving every applicant a clear explanation and a meaningful path to correct or contest the information used."**

The difference wasn't cosmetic. A purpose written around the person creates different product requirements than a purpose written around the process.

You can apply the same test to ordinary personal use:

> ***"Examine the AI-supported decision below from the per-***

spective of the person most affected by it. What would they need to know, consent to, correct, refuse, or appeal in order to retain dignity and meaningful choice? Identify any convenience for me that creates hidden cost or surveillance for them."

That question may change how a manager uses sentiment analysis, how a parent configures a learning tool, how a creator handles another person's likeness, or how a household shares location data. Ethics isn't only a policy written by institutions. It's a habit of noticing when your convenience is being purchased with someone else's agency.

Turn Principles into Practice

Principles matter most when the deadline is close, the demo is impressive, and asking another question feels inconvenient.

Evelyn's team built a lifecycle around that moment. Before development, the owner had to state the purpose in plain language, name the people affected, and explain why AI was needed. A problem that could be solved with a simple rule or clearer process didn't become more responsible merely because a model was available.

During design, the team defined what the system could recommend, what it could never decide alone, what information it required, and what explanation would accompany the result. High-impact uses faced a higher bar and required qualified legal, technical, and domain review appropriate to the location and industry. Laws and standards change. A general framework can't replace current professional guidance about a real deployment.

Before launch, the team tested more than the average case. It examined incomplete records, conflicting sources, unusual income patterns, language differences, accessibility needs, attempts to manipulate the system, and situations in which the correct response was "I don't know." People outside the development team were invited to challenge the assumptions. The pilot remained small enough that learning didn't require widespread harm.

In production, the work continued. The organization monitored errors, uneven outcomes, complaints, reversals, security incidents, and whether employees were relying on the recommendation more heavily than intended. A named owner had the authority to pause the system. A clear channel allowed customers and employees to report concerns without first proving they understood the technology.

The team used the RIGHTS framework to keep the essential commitments close:

Reasonable purpose: Is AI appropriate for this problem, and is the benefit worth the risk?

Informed permission: Do people understand what's collected, how it's used, and what choices they have?

Guardrails: Are permissions limited, high-impact actions reviewed, and foreseeable misuse tested?

Human impact tests: Have outcomes been examined across

the people and conditions the system may serve poorly?

Transparency: Can an affected person understand, correct, and challenge the result?

Sharing gains: Do the benefits of automation reach the customers, workers, and communities carrying its effects?

The framework was a memory aid, not a certificate. A team could complete every line mechanically and still miss the person in front of it. Evelyn added one final prompt to every launch review:

> ***"Assume this system has been operating for one year and has caused a serious but plausible harm. Write the incident from the affected person's point of view. Then identify which assumption, control, warning, or appeal process failed, what evidence would've revealed the problem earlier, who should've acted, and what repair would be owed."***

Imagining failure didn't mean expecting failure. It turned vague confidence into specific preparation.

Nia's application became one of the cases the team reviewed. The model had interpreted irregular but documented business income conservatively and the original process hadn't made the uncertainty clear. A qualified reviewer examined the underlying records under the lender's policies. Whether the final lending decision changed wasn't the only ethical issue. The process now had to give Nia a specific explanation, identify what information mattered, and provide a real way to correct or supplement the record.

Evelyn called her.

The conversation wasn't easy. Nia was frustrated, and a better explanation didn't erase the weeks she'd spent feeling judged by a number she couldn't see. Evelyn listened, explained what the review had found, described the next step, and told her what the team was changing. Accountability required more than improving the model. It required facing the person who'd carried the failure.

That's the hinge beneath every AI system: human choice remains present even when automation makes it less visible. People decide what to build, what to collect, what to optimize, what to explain, and what to do when the result harms someone. Millions of those decisions become the culture in which everyone has to live.

Be the person who asks "according to whom?" when a system calls itself fair. Be the manager who measures trust as well as speed. Be the parent who values adaptive help and still protects a child's privacy. Be the creator who makes the process clear. Be the customer who asks for an explanation and the leader who ensures there's one to give.

The models will continue improving. The systems will become easier to use and harder to see. The question of what they're allowed to serve won't answer itself.

When dignity remains in the loop, AI can extend human ability without turning human beings into inputs. When dignity leaves the loop, efficiency becomes a poor defense.

The future will be shaped not only by what machines learn to do, but by what people choose to remain responsible for.

18

What's Next in AI

Imagine an ordinary Tuesday a few years from now. You wake to a short briefing that's already compared your calendar, the weather, traffic, a school schedule change, and the work you said mattered most this week. It hasn't sent a message or moved an appointment. It's prepared three options, explained the tradeoffs, and is waiting for you to decide.

Later that morning, you give the same system a larger goal: prepare for a difficult client meeting. It gathers the approved files, traces the decisions that produced the current problem, checks the latest numbers, builds a first version of the analysis, and identifies the questions it can't answer from the evidence. You correct one assumption, approve two actions, and reject a third. By the time the meeting begins, the system has carried much of the coordination. **You still carry the consequence.**

Nothing in that scene requires a conscious machine or a robot waiting beside the coffee maker. It requires something more practical: AI that can perceive different kinds of information, remember relevant context, use tools, pursue a multi-step goal, and stop at the boundaries a person has set. Pieces of that future already exist. What comes next is the steady connection of those pieces.

That's why the AI frontier is difficult to describe with a single prediction. Progress is real, but uneven. A model may solve an advanced mathematics problem and then stumble over a simple visual detail. An agent may complete a complicated digital workflow and still misunderstand one ambiguous instruction. The frontier isn't a clean line separating the possible from the impossible. It's a moving, jagged edge where astonishing ability and ordinary failure live side by side.

The useful question isn't whether every forecast will arrive on schedule. It's which capabilities are becoming durable enough to change how people live and work—and what those capabilities will require from the people who use them.

From Answers to Action

The first wave of generative AI taught millions of people to ask a machine for an answer, an image, a summary, or a draft. The next wave is teaching machines to carry a process. A chatbot waits for the next instruction. **An agent works toward an outcome.** It can break a goal into steps, choose among approved tools, examine the result of one action, and use that result to decide what should happen next.

That shift sounds technical until it enters a normal week. Instead of asking for a travel checklist, you may ask an agent to compare routes, respect a budget, flag cancellation risk, assemble an itinerary, and hold every purchase for approval. Instead of requesting a meeting summary, you may ask it to identify commitments, update a project plan, draft follow-ups, and return unresolved decisions to their owners. Instead of asking what a health-insurance letter means, you may ask it to organize the documents, explain the terms in plain language, identify missing information, and prepare questions for a qualified professional.

The value comes from continuity. Today, people spend enormous energy reintroducing the same context: the goal, the constraints, the history, the people involved, and the definition of done. Future systems will become more useful as they can retain the right context across time without requiring a person to rebuild the entire story at every turn.

That context may become deeply personal. An assistant could learn that you prefer direct explanations, that your family protects Wednesday evenings, that a certain account is never used for discretionary spending, or that health questions must be grounded in sources and reviewed with a clinician. It may understand your files, routines, communication style, and recurring responsibilities well enough to notice a conflict before you do.

Personalization can feel like relief because it reduces the tax of explaining yourself. It can also become surveillance disguised as convenience. The more context a system holds, the more important it becomes to know where that information lives, how long it remains, who can use it, and whether you can inspect, correct, or delete it. **A useful memory should serve the person being remembered.**

One important part of this frontier will happen closer to you—on phones, computers, vehicles, appliances, and other devices rather than only in distant data centers. Smaller models running locally can respond with less delay, work when a connection is unreliable, and keep some sensitive tasks on the device. The most capable systems may combine both approaches, handling ordinary personal context locally and reaching for larger models only when a task truly requires them.

The interface will also become less dependent on typing. Multimodal systems can already work across text, speech, images, documents, charts, and software screens. As those abilities improve, AI will be able to follow a conversation while examining the spreadsheet on the screen, notice that a diagram conflicts with the written plan, or help someone understand an unfamiliar machine by looking through a camera. The interaction will feel less like filling out a form and more like sharing attention.

Eventually, a single visible assistant may coordinate several specialized systems behind the scenes: one checks sources, another works through numbers, another controls a calendar, and another tests the final result against policy. You may not care which model completed each step any more than you care which server delivered a webpage. What will matter is whether the system used the right information, respected its permissions, showed its work, and produced a result you can trust.

This is where excitement must mature into design. Greater autonomy is valuable only when it's paired with **clear authority**. An agent that can draft a message is useful. An agent that can send it, spend money, sign an agreement, change a medical record, or act on another person's behalf crosses into a different level of consequence. The system shouldn't receive that authority merely because it has the technical ability to use it.

A practical way to prepare is to build a permission ladder. At the bottom, AI may read approved information and suggest. On the next level, it may draft and organize. Above that, it may act inside narrow rules and reversible limits. At the top are consequential actions that require an informed person to approve, explain, and remain accountable. The ladder will differ by task, but the principle travels: **permission should rise more slowly than capability.**

> ***"Design a permission ladder for the recurring workflow below. Separate what AI may observe, suggest, draft, and execute. Identify which actions are reversible, which could affect another person, which involve money or sensitive data, and which require my approval. Include a visible activity log, a spending or action limit, and a clear stop condition when the system is uncertain."***

The goal isn't to keep AI weak. It's to make powerful systems dependable enough to deserve a role in real life. The best agent may not be the one that acts most often. It may be the one that knows when to pause, asks the question that prevents a costly mistake, and makes it easy for a human being to understand exactly what happened.

As agents become more capable, your role moves from producing every step to designing the conditions under which the work happens. You become less like a person giving isolated commands and more like an operator setting intent, permissions, evidence requirements, and review points. That isn't the disappearance of human work. It's a shift toward work that carries more judgment.

When Intelligence Leaves the Screen

The digital frontier is only part of the story. AI becomes more consequential when it connects language and reasoning to scientific instruments, simulations, sensors, and machines that can affect the physical world.

In science, the change is already moving beyond faster literature reviews. AI systems can help researchers connect ideas across fields, search enormous spaces of possible molecules or materials, interpret complex data, write and test analysis code, and propose hypotheses worth examining. A researcher may use AI to generate a larger set of plausible explanations, eliminate weak paths earlier, and design the next experiment with more information than a human team could gather manually in the same amount of time.

Imagine a laboratory investigating a disease that behaves differently across patient groups. The AI doesn't discover a cure in one dramatic moment. It helps the team compare papers across disciplines, examine patterns in genomic and clinical data, model several mechanisms, and suggest experiments that could distinguish one explanation from another. Scientists challenge the assumptions, run the experiments, and reject most of the ideas. One survives. The breakthrough, if it comes, belongs to the human–AI system: machine breadth joined to human expertise, physical evidence, and the discipline to prove what's true.

That distinction matters. A plausible hypothesis isn't a discovery. A simulated result isn't a safe treatment. A generated proof isn't complete until it's been checked, and a confident citation isn't a source until someone verifies it exists. AI can accelerate the search and sometimes contribute a valuable idea, but science advances because evidence can withstand challenge. **Speed can shorten the path to a test; it can't replace the test.**

Some of the most useful systems won't be the largest general models. They'll be specialized intelligence connected to the tools of a field: a microscope, a weather model, a protein database, a telescope, an engineering simulator, or a manufacturing line. General reasoning can help connect the pieces; specialized systems supply the precision and feedback that real discovery requires.

Robotics brings the same progression into motion. Traditional robots are extraordinary inside carefully designed conditions. They repeat known actions with speed and precision, but they often struggle when an object moves, a room changes, or an instruction contains ambiguity. Newer embodied systems are learning to combine vision, language, spatial reasoning, planning, and movement so they can adapt rather than merely replay a script.

The first broad impact is likely to be uneven and practical. Robots may inspect dangerous infrastructure, move materials through warehouses, support laboratory work, tend crops, assist in disaster zones, or handle repetitive tasks in environments built around safety. In hospitals and elder care, carefully designed machines may transport supplies, monitor routine conditions, or help with mobility. The promise isn't a metal person who replaces every human role. It's physical assistance in places where risk, repetition, distance, or labor shortages make help valuable.

Homes will present the hardest standard because a home isn't a factory. It contains children, pets, private conversations, uneven floors, changing routines, and people who haven't agreed to become training data. A household robot that can lift, carry, observe, and learn may be genuinely useful to an older adult or a person with a disability. It may also create new forms of dependence, surveillance, or physical risk. The question won't be only whether the machine can complete the task. It'll be whether it can do so while preserving safety, privacy, dignity, and the person's ability to say no.

Before more intelligence moves into the physical world, much of it'll work through simulation. Digital models of factories, buildings, supply chains, vehicles, storms, and biological systems allow people to test choices before making them real. AI can explore more scenarios, identify failure patterns, and surface tradeoffs that'd be expensive or dangerous to discover through trial and error. The frontier here isn't prediction without uncertainty. It's better rehearsal.

Rehearsal changes the quality of preparation. A city can explore how a flood might affect roads and hospitals. A manufacturer can test a line change before interrupting production. A doctor can review a model's possible interpretation while remaining responsible for the clinical decision. A family can compare financial scenarios without confusing a forecast for a guarantee. In every case, simulation expands the set of questions people can afford to ask.

Once AI acts through software, instruments, or machines, mistakes no longer remain inside a chat window. They can move money, change access, damage equipment, expose private information, or place a body at risk. Physical and scientific capability therefore demands layered safety: limited permissions, independent checks, secure systems, human stop controls, testing under unusual conditions, and named people who own the outcome.

> ***"Map how the following field could change as AI gains three abilities: persistent context, multi-step action, and access to scientific or physical tools. Separate what's possible now, what appears plausible over the next few years, and what remains speculative. For each possibility, identify the evidence, human expertise, permission, safety control, and real-world validation that'd still be required."***

You don't need to work in a laboratory or own a robot to use this way of thinking. The deeper lesson is to treat AI as a system that can help explore possibilities while reality keeps the final vote. Ask it to model, compare, rehearse, and challenge. Then bring the result back to the evidence, the expert, the customer, the physical constraint, or the person whose life will carry the decision.

The most important advances may arrive quietly: an experiment that fails sooner and saves a year, an inspection completed without placing a worker in danger, a treatment path tested more precisely, or an assistive device that returns a small piece of independence. The frontier becomes meaningful when capability leaves the demonstration and enters a life with care.

The Frontier That Still Belongs to Us

If intelligence becomes more capable, less expensive, and easier to direct, many forms of expertise will become available to more people. A small company may analyze information that once required a large department. A student may receive patient explanation in a language no local tutor speaks. A creator may test an idea before asking for funding. A clinician or scientist may examine possibilities that were previously too numerous to explore. **Abundant intelligence can widen the doorway to action.**

It won't distribute opportunity automatically. Access to tools isn't the same as access to reliable data, education, time, capital, connectivity, or the confidence to challenge a machine. Some organizations will use AI to expand people's capability; others will use it mainly to reduce labor or increase control. Some communities will help shape the systems that affect them; others may receive decisions they can't see or appeal. The future can become more capable and less fair at the same time.

Work will change through tasks before it changes through titles. Research, analysis, scheduling, coding, design, service, and documentation will contain larger automated sections. Some roles will shrink, some will grow, and new combinations will appear. The person who once produced every first draft may supervise a stream of drafts. The manager who once waited for a report may receive a live explanation and need to decide sooner. The entrepreneur may coordinate capabilities that once required a full staff.

That leverage can create extraordinary independence. One person with domain knowledge, a trusted network, and well-designed AI systems may build a business, conduct research, create a film, teach a course, or serve a community at a scale that'd have been impossible a decade earlier. But the same leverage means one person can spread an error, invade privacy, imitate another person's work, or automate a harmful decision farther than before. **Reach and responsibility rise together.**

The strongest protection against the downside isn't refusing to learn. It's learning deeply enough to keep judgment alive. When people treat AI as magic, they surrender too much. When they treat it as worthless because it sometimes fails, they surrender opportunity. Maturity lives between those reactions: understand what the system is good at, recognize where it's brittle, and design the relationship so its strengths don't hide its limits.

Five human capabilities will become more valuable as the tools improve. The first is **problem framing**: deciding which problem deserves attention and what a good outcome means. A system can optimize the goal it receives; it can't guarantee that the goal is wise.

The second is **verification**: tracing a claim to evidence, checking the math, testing the output against reality, and knowing when a qualified professional belongs in the loop. In a world of fluent answers, the willingness to slow down for truth becomes a competitive and moral advantage.

The third is **orchestration**: giving several forms of intelligence a shared purpose, clear roles, useful context, and stop conditions. This is more than prompt writing. It's the ability to design a process in which the right work reaches the right tool and the consequential judgment returns to the right person.

The fourth is **taste and trust**. When competent content becomes plentiful, people will value those who can recognize what's worth keeping, communicate with a real voice, honor a promise, and understand the human being on the other side of the output. Relationships won't become less important because production becomes easier. They may become the clearest signal that the work is real.

The fifth is **moral courage**: the ability to refuse an efficient choice that violates dignity, to disclose uncertainty when confidence would be more convenient, to protect someone's information when exploiting it'd be profitable, and to remain accountable when a machine participated in the mistake. No model can carry that responsibility away from you.

These capabilities aren't preparation for a distant decade. You can practice them now. Choose one workflow that matters. Give AI a bounded role. Ask it to show evidence. Record what improved and what failed. Protect one part of the work for human presence. Review the result with the person affected. Small experiments build discernment faster than passive fascination.

"Help me create a 90-day frontier practice for my life or work. Choose one recurring workflow to improve, one durable human skill to deepen, and one ethical boundary to protect. Design a small weekly experiment, define evidence of real value, identify what I must verify myself, and include a review question that tests whether AI is increasing my agency without reducing someone else's."

For leaders, preparation also means making room for other people to learn. Don't introduce AI only as a cost target and then ask employees to feel inspired. Explain the purpose, involve the people who understand the work, invest in new skills, measure quality as well as speed, and make it safe to report a failure. If the gains are real, decide how customers, employees, and communities will share them.

For parents and educators, the task isn't to keep young people untouched by AI. It's to help them become hard to fool and eager to think. Let them use the tool, then ask where the answer came from, what it missed, how they'd explain it without the screen, and which part of the work should remain theirs. Teach creation and skepticism together.

Don't prepare for the frontier by memorizing product names or betting your identity on one forecast. Products will be renamed. Interfaces will disappear. Predictions will be revised. Build knowledge in a field you care about, learn how to collaborate with intelligent systems, strengthen the habits that protect your body and attention, and decide your values before convenience puts them under pressure.

A frontier isn't simply the place where technology becomes more powerful. It's the place where familiar maps stop being enough. That can be frightening, but it's also where new possibility enters. There are diseases still waiting for better questions, businesses waiting for someone who understands the overlooked problem, stories waiting for a creator who can finally afford the first version, and hours of human attention waiting to be returned to people who need it.

The next era of AI will be shaped through millions of ordinary choices: what we delegate, what we verify, what we refuse, what we disclose, whom we include, and what we do with the time and power we gain. The machines will become more capable. That part is already underway. **The human frontier is deciding what that capability is for.**

Conclusion: Beyond Human Limits—Your Next Advantage

Tomorrow morning, almost nothing around you'll look different because you finished this book. Your calendar may still be crowded. The message you've been avoiding will still be waiting. The idea you keep saying you'll start "when things calm down" won't have built itself overnight.

But you may see those things differently now. The crowded calendar isn't only a fact; it's a system that can be redesigned. The difficult message isn't only a source of anxiety; it's a moment where AI can help you prepare while your judgment and voice remain in charge. The delayed idea isn't only a dream; it's a series of smaller questions, drafts, experiments, and decisions that no longer have to begin from a blank page.

That shift is the promise running through this entire book. AI isn't valuable because it can make you resemble a machine. It's valuable because it can carry some of the friction that's kept you from doing work, learning skills, creating things, caring for people, and making decisions at the level you know is possible.

We began with productivity, but productivity was never the final destination. We moved through decisions, mindset, health, attention, emotional intelligence, business, wealth, learning, creativity, home, ethics, and the frontier ahead because your life isn't divided into neat technical categories. It's one life. The same intelligence that helps organize a workday can help you understand a financial question, prepare for a medical conversation, teach yourself a skill, test a business idea, or protect an hour for your family. The real advantage appears when those pieces serve a coherent human purpose.

The Life You Can Build

Human limits aren't personal failures. Attention is finite. Memory is imperfect. Energy changes. Time passes at the same rate for everyone. Fear makes a first step feel larger than it is, and complexity can bury a good decision beneath so much information that we postpone it. For most of history, people had to carry those limits with only the tools, knowledge, and help close enough to reach.

AI changes the amount of leverage available to an ordinary person. It can become a working memory for a project, a patient tutor, a first-draft partner, a research assistant, a simulator of choices, a pattern finder, or a coordinator that keeps routine work from consuming the day. It can help one person reach capabilities that once belonged only to a large team or a specialist with time to spare.

Think about what that makes possible. The business idea you've carried for years can become a simple offer you test with real people. The subject that once felt inaccessible can be explained three different ways until one finally opens the door. A pile of household information can become a plan the whole family understands. A confusing set of expenses can become questions you take to a qualified advisor. A creative impulse can become a rough page, a sketch, a melody, or a prototype before doubt has time to close it down.

The point isn't to fill every recovered minute with more output. A life optimized only for volume becomes another machine you have to serve. The deeper opportunity is to exchange friction for intention: less time assembling information, more time understanding it; less energy repeating routine coordination, more energy making the call that matters; fewer blank-page battles, more honest revision; less administrative noise, more presence.

You aren't too late. You don't need to become a programmer, follow every product release, or wait for an employer to hand you permission. The entry point is already in front of you: one question you need to understand, one burden you need to lighten, one idea you want to make real, or one part of your life that deserves a better system.

This is the democratization of intelligence in its most personal form. It doesn't mean everyone receives the same outcome. It means more people can reach a capable first partner, explore a possibility, and arrive at the starting line with something more than hope. The old gatekeepers don't all disappear, and expertise doesn't stop mattering. But the cost of beginning falls.

There are projects inside you that may have been waiting for exactly that change. Not because AI contains your idea, but because it can help remove enough resistance for you to meet the idea with your own courage. The future should make you hungry—not for endless novelty, but for the chance to attempt work that used to feel too complicated, too expensive, too lonely, or too far outside your training.

Still, freedom from friction doesn't tell you what deserves your freedom. Recovered time can become deeper work or more distraction. Greater reach can become service or self-importance. Faster creation can reveal your voice or bury it beneath imitation. **Leverage creates possibility; purpose gives it direction.**

The Responsibility That Comes with Reach

Every promise in this book has a shadow. AI can help you decide faster, which means you can spread a weak assumption faster. It can personalize support, which means it may require information that deserves protection. It can make creation accessible, which means it can also make deception cheap. It can give one person the reach of a team, which means one person's carelessness can travel farther than before.

Responsibility isn't the price we pay for progress after the exciting part is over. **Responsibility is how progress becomes**

worth having. It's the steering that allows speed to serve a destination. It's the trust that lets other people accept help without surrendering their dignity. It's the discipline that separates useful amplification from careless scale.

Never hand off the consequence merely because you handed off the task. If AI drafts the recommendation, you're still responsible for the evidence behind it. If it organizes a medical or financial question, a qualified person still belongs in the consequential decision. If it writes in your voice, you're responsible for whether the message is honest. If an agent acts with your permission, the effect of that action still carries your name.

Protect privacy before convenience makes the choice feel automatic. Verify a claim before fluency makes it feel true. Ask whose work shaped the output and give credit where it's due. Explain when AI materially participated in something another person is being asked to trust. Preserve a way for people to question, correct, refuse, or appeal decisions that affect them. When the stakes rise, slow the system down enough for dignity to remain visible.

These choices aren't reserved for technology companies or lawmakers. You make them when you upload a family document, use a child's information, summarize an employee's performance, generate a person's likeness, accept a market explanation, or let an automated recommendation shape who receives your time. Ethics becomes culture through ordinary behavior long before it becomes policy.

Leaders carry a wider version of the same obligation. When you introduce AI into someone else's job, your decision changes more than a workflow. It changes what's measured, which skills are valued, how closely a person is watched, whether mistakes can be challenged, and who receives the benefit of the new efficiency. Bring people into the design. Train them for the changed work. Share the gains. Name the person who can stop the system when it stops serving its purpose.

Responsibility shouldn't leave you frozen. Fear can become another excuse to avoid learning, leaving the most consequential choices to people who never paused to ask the questions you'd have asked. Enter the work. Learn enough to recognize both possibility and harm. Build the guardrail, run the small test, listen to the person affected, and improve the system before scale hardens a poor choice.

The person you become in this transition matters more than the number of tools you master. Be curious without becoming gullible. Be ambitious without treating people as inputs. Be efficient without confusing speed with wisdom. Be willing to change your mind when the evidence changes. Keep a part of your life beyond measurement so you remember that not everything valuable needs to become a metric.

AI can extend your reach beyond familiar human limits. It can't extend you beyond human responsibility. **The farther the reach, the more carefully you must choose what your hand is doing.**

Begin Before You Feel Ready

Change rarely begins with a grand transformation. It begins when you stop accepting one recurring frustration as inevitable. It begins when you let a tool help with the first draft and discover that your real value was never the blank page. It begins when you question a routine, protect a boundary, or test an idea before your doubts can turn it back into someday.

Start with three commitments.

Reduce one friction. Choose a repetitive task that consumes attention but doesn't deserve your best judgment. Let AI help organize, summarize, compare, schedule, or prepare. Keep the lane narrow enough that you can see what changed.

Advance one ambition. Return to something meaningful that's been stalled by complexity or uncertainty. Ask for a map, a first experiment, a critique, or a teachable next step. Use the machine to make movement cheaper, then bring your effort to the part that requires courage.

Protect one boundary. Decide in advance what remains yours: a sensitive message, a financial decision, a child's privacy, the final creative choice, a daily hour without a screen, or the obligation to verify before you act. A boundary isn't resistance to the future. It's evidence that you know what the future is for.

> ***"Help me design my first seven days of intentional AI use. Begin with one recurring friction I want to reduce, one meaningful goal I want to advance, and one boundary I***

refuse to cross. Propose one small experiment per day, identify what I must verify, and end each day with one question that keeps the final judgment and responsibility mine."

Treat the response as a draft, not an order. Make it fit your actual life. Then do the first small thing before you collect another list of possibilities. The skill that matters isn't knowing everything AI can do. It's turning one useful possibility into a responsible practice.

At the end of the week, ask what became lighter, what became better, and what became less human. Keep the first two. Repair the third. At the end of the month, look for evidence: time returned, a decision improved, a skill deepened, a project moved, a conversation handled with more care. If the tool created more work than it removed, change it or let it go.

When an experiment fails, don't treat the failure as a verdict on the entire future. Learn whether the problem was the tool, the instructions, the data, the workflow, or the task you should never have delegated. When an experiment succeeds, don't stop at personal advantage. Document the practice, teach someone else, and ask how the benefit can travel without the harm traveling with it.

There'll be no final moment when you're fully caught up. The tools will keep changing. That isn't a flaw in your preparation; it's the condition of the age. Your durable advantage is the ability to learn, test, judge, and adapt without handing your values to the newest interface.

You won't become irreplaceable by competing with a machine at the tasks machines do best. You become difficult to replace for better reasons: because people trust your judgment, because you understand the problem beneath the request, because you can tell the truth when an easier answer is available, because your taste gives direction to abundance, and because you know how to make another human being feel seen.

The future described in these pages isn't waiting on a distant stage. It's arriving inside your next email, your next question, your next purchase, your next meeting, your next hour with someone you love. Each moment offers the same choice: use intelligence to create more noise, or use it to make room for what matters.

Let AI carry more information, but not your integrity. Let it accelerate the draft, but not rush the truth. Let it expand what you can attempt, but never shrink the people your decisions affect.

You aren't being replaced. You're being asked to become more intentional about what you amplify.

There's work ahead that only you can choose to begin. There are limits you can finally move beyond and responsibilities you must refuse to leave behind. That's the invitation of this moment: not humans versus machines, and not machines over humans, but human beings with greater reach and a clearer obligation to use it well.

Close the book. Choose the first change. Build the future on purpose.

AI Toolkit: Practical Tools for Every Area of Life

Tool snapshot: August 2026. Product names, features, availability, and pricing change quickly. Verify current terms and privacy settings before you adopt anything.

A useful stack isn't the one with the most logos. It's the smallest set of tools that removes real friction while leaving judgment, relationships, and high-stakes decisions in human hands.

How to Choose a Tool

- **Start with friction.** Name the recurring task or decision before you shop for an app.
- **Prefer tools already near your work.** Built-in AI often creates less switching and fewer new data connections.

- **Check the data path.** Know what the tool can read, what it stores, whether data trains models, and how to delete it.

- **Require review where stakes rise.** Money, health, employment, legal rights, safety, and external messages need a qualified human.

- **Review quarterly.** Keep what saves time or improves quality; remove what creates maintenance, distraction, or risk.

1. Productivity and Work

- **General assistants—ChatGPT, Claude, or Gemini.** Use for outlining, drafting, synthesis, planning, voice conversations, and critique. Choose one primary assistant and learn it well.

- **Work suites—Microsoft 365 Copilot, Gemini in Workspace, or Notion AI.** Use AI where your documents, email, meetings, and team knowledge already live, subject to your organization's policies.

- **Meeting support—Granola or Otter Meeting Agent.** Capture notes, summaries, and action items only with appropriate notice and consent. Granola emphasizes bot-free personal notes; Otter emphasizes automated transcripts and meeting workflows.

- **Scheduling—Reclaim or Motion.** Protect focus time, place tasks on the calendar, and reschedule around real constraints.

- **Automation—Zapier Agents or Make AI Agents.** Connect approved apps and automate repeatable workflows. Add human approval before sending messages, spending money, or changing consequential records.

2. Health and Fitness

- **Oura Advisor or WHOOP Coach.** Use wearable data to explore patterns in sleep, activity, strain, and recovery—not to diagnose illness or override symptoms.

- **Eight Sleep Autopilot.** Automate sleep-temperature adjustments for people who already use the Eight Sleep system.

- **ZOE.** Use food logging and personalized nutrition insights as a source of questions and experiments, not medical certainty.

Health tools can be wrong, incomplete, or overly confident. Confirm important changes with a qualified clinician, especially when symptoms, medication, pregnancy, injury, or a medical condition is involved.

3. Money and Financial Planning

- **Monarch AI Assistant.** Ask natural-language questions about your connected financial picture, changes, goals, and spending patterns.

- **Copilot Money.** Use learned transaction categorization, spending views, and its developing Money Assistant to keep financial data cleaner and easier to review.

- **Betterment or Wealthfront.** Consider automated investing and rebalancing if they fit your situation, while recognizing that automation isn't a guarantee and fees and tax effects matter.

Don't use a general chatbot for stock picks, tax filings, or personalized financial decisions. Avoid uploading unredacted statements unless you've deliberately reviewed the tool's security, retention, and training policies.

4. Learning and Research

- **Khanmigo.** Use guided tutoring that asks questions and supports learning rather than simply handing over an answer.

- **Duolingo and Duolingo Max.** Use daily language practice, explanations, roleplay, and supported video conversations where available.

- **Gemini Notebook (formerly NotebookLM).** Build a source-grounded notebook from material you select, then ask questions, compare sources, and create study aids.

- **Elicit.** Search, summarize, and extract information from scientific literature while opening the papers and checking the evidence yourself.

- **Perplexity.** Begin web research with inline citations, then inspect the original sources before relying on a claim.

5. Creativity and Expression

- **ChatGPT Images or Midjourney.** Explore concepts, compositions, edits, and visual directions. Use the output as material to shape, not a substitute for taste.
- **Canva AI or Adobe Firefly.** Move from idea to editable graphics, presentations, images, video, or audio inside a broader creative workspace.
- **Runway.** Generate, edit, extend, and prototype video when moving images are central to the idea.
- **Suno.** Prototype songs and musical ideas while checking current licensing terms before commercial release.

For public or commercial work, verify usage rights, disclose material AI involvement when it matters, obtain consent for real people's likenesses or voices, and avoid imitating living creators in ways you wouldn't defend openly.

6. Home, Relationships, and Daily Life

- **Voice mode in your general assistant.** Brainstorm meals, rehearse a difficult conversation, translate a phrase, or turn a spoken brain-dump into a plan.
- **Reclaim, Motion, or a shared calendar.** Coordinate

the movable parts of a household without treating every family member like an employee.

- **Notion AI.** Maintain a shared home page for plans, checklists, travel details, or projects when everyone agrees on what belongs there.

- **Tiimo or visual-timer tools.** Support transitions and executive function with visible, compassionate cues rather than surveillance.

Automate coordination, not connection. Don't record family, school, medical, or co-parenting conversations without appropriate consent, and keep intimate household data out of tools that don't need it.

7. Business and Entrepreneurship

- **HubSpot Breeze and Agent Hub.** Create content, find information, automate go-to-market workflows, and manage specialized agents inside HubSpot.

- **Salesforce Agentforce and Einstein.** Use CRM-grounded predictions, content, and agents with clear permissions, explanations, and human escalation.

- **Shopify Sidekick and Shopify Magic.** Ask questions about store data, draft commerce content, and handle supported administrative tasks inside Shopify.

- **Wix Harmony or Wix AI Website Builder.** Create

a business-ready site conversationally, then edit the copy, design, accessibility, and claims before publishing.

- **Harvey.** Support legal research, contract analysis, and professional workflows in qualified legal teams—not as a substitute for an attorney.

Recommended AI Starter Stack

Don't install all of these at once. Choose two or three categories tied to a real need, use them for a month, and add another only when the value is clear.

1. One General Assistant

Choose ChatGPT, Claude, or Gemini as your main thinking partner. Use it for plans, drafts, critique, learning, and reflection; keep sensitive data out unless the account and policy are appropriate.

2. One Work-Suite Assistant

Choose Microsoft 365 Copilot, Gemini in Workspace, or Notion AI based on where your work already lives. Integration is more useful than novelty.

3. One Research Tool

Choose Gemini Notebook for work grounded in your own sources, Perplexity for cited web exploration, or Elicit for scientific literature. In every case, open the sources.

4. One Meeting Tool

Choose Granola when you want a personal, bot-free AI notepad, or Otter when you want automated transcription and team meeting workflows. Get consent first.

5. One Time Tool

Choose Reclaim or Motion to protect focus blocks and place tasks on a realistic calendar. The tool can negotiate time; you still decide what deserves it.

6. One Automation Tool

Choose Zapier or Make only after you understand the manual process. Start read-only where possible, log what happens, and require approval for consequential actions.

7. One Creative Tool

Choose Canva AI for editable everyday design, ChatGPT Images for conversational generation and editing, or Adobe Firefly or Midjourney when visual exploration is central to your work.

8. One Specialist Tool

Choose a specialist only when a current goal justifies it: Oura or WHOOP for wellness patterns, Monarch or Copilot Money for financial organization, or Wix or Shopify when you're actively launching a business.

The strongest stack is the one you can explain, maintain, and turn off. Tools should make your life feel more intentional—not make you feel responsible for another collection of machines.

About the Author

Anthony Case has spent more than 25 years helping organizations navigate the ever-changing world of technology, strategy, and innovation. Throughout his career, he has worked with major corporations to solve complex business challenges, improve performance, and transform bold ideas into practical results.

Believing that artificial intelligence is one of the most significant technological shifts of our lifetime, Anthony is passionate about helping others move beyond the hype and discover how AI can become a practical partner in everyday life. His goal is to make complex technology approachable, empowering readers to think bigger, work smarter, and embrace the opportunities of the future with confidence.

When he's not exploring the latest advances in AI or writing about the future of technology, Anthony's greatest joy is spending time with his wife and children. As the proud father of four daughters, he understands that the most meaningful innovations aren't just the ones that change how we work—they're the ones that give us more time to invest in the people we love.

Through his writing, Anthony hopes to inspire readers to see artificial intelligence not as something to fear, but as a powerful tool that can help us create, learn, solve problems, and build a better future.